Forensics Demystified

Demystified Series

Forensics Demystified

Barry A. J. Fisher, MS, MBA
Crime Laboratory Director
Los Angeles County Sheriff's Department

David R. Fisher, BS, D-ABC
Criminalist III
Department of Forensic Biology
New York City Office of Chief Medical Examiner

Jason Kolowski, MS, D-ABC
Criminalist III
Department of Forensic Biology
New York City Office of Chief Medical Examiner

with a Foreword by
Dr. Peter Pizzola
Crime Laboratory Director
New York City Police Department

McGraw-Hill

New York Chicago San Francisco Lisbon London
Madrid Mexico City Milan New Delhi San Juan
Seoul Singapore Sydney Toronto

McGraw-Hill books are available at special quantity discounts to use as premiums and sales promotions, or for use in corporate training programs. For more information, please write to the Director of Special Sales, Professional Publishing, McGraw-Hill, Two Penn Plaza, New York, NY 10121-2298. Or contact your local bookstore.

Forensics Demystified

1234567890 DOC DOC 019876

ISBN -13: 978-0-07-145430-8
ISBN -10: 0-07-145430-6

Sponsoring Editor Judy Bass	**Production Supervisor** Jean Bodeaux
Editorial Supervisor Janet Walden	**Composition** TechBooks
Project Manager Ronald D'Souza	**Illustration** Irene Wong
Copy Editor William McManus	**Cover Series Design** Margaret Webster-Shapiro
Proofreader Sunita Dogra	**Cover Illustration** Lance Lekander
Indexer Karin Arrigoni	

This book is dedicated to the men and women who lost their lives on September 11, 2001, and to the dedicated employees of the Department of Forensic Biology at the New York City Office of Chief Medical Examiner, who pushed the boundary of forensic science in their efforts to identify and return to their loved ones, those who were murdered on that day.

ABOUT THE AUTHORS

Barry Fisher is Crime Laboratory Director for the Los Angeles County Sheriff's Department and has worked in various sections within the lab since starting in 1969. He is a distinguished fellow and past president of the American Academy of Forensic Sciences and served as president of the American Society of Crime Laboratory Directors (ASCLD) and the International Association of Forensic Sciences. He is a past chair of the ASCLD Laboratory Accreditation Board. He is a member of several other organizations, including the International Association for Identification, International Association of Chiefs of Police, the California Association of Criminalists, and others. He holds a BS in chemistry from City College of New York, an MS in organic chemistry from Purdue University, and an MBA from California State University, Northridge. He is author of the popular textbook, *Techniques of Crime Scene Investigation*, now in its 7th edition, published by CRC Press.

David Fisher is a Criminalist, Level III in the Department of Forensic Biology at the Office of Chief Medical Examiner in New York City. He is a Diplomate in the American Board of Criminalistics and is a member of the Northeastern Association of Forensic Scientists and an associate member of the American Academy of Forensic Sciences. David was recently deployed to the Gulf region with the Disaster Mortuary Operational Response Team (DMORT), which is part of the National Disaster Medical System under FEMA to aid in the identification of the victims of Hurricane Katrina. He holds a BS in biochemistry and cell biology from the University of California, San Diego and is currently finishing his MS degree in forensic science at the John Jay College of Criminal Justice at the City University of New York.

Jason Kolowski is a Criminalist, Level III in the mitochondrial DNA laboratory within the Department of Forensic Biology at the Office of Chief Medical Examiner in New York City. He is also a Diplomate in the American Board of Criminalistics, and a member of the American Academy of Forensic Sciences and the Northeastern Association of Forensic Scientists. Jason is a member of the DMORT Region II team under FEMA. Jason attended Cornell College where he earned a dual BA degree in biochemistry/molecular biology and philosophy. Jason earned his master's degree in forensic science in 2003 from the John Jay College of Criminal Justice, at the City University of New York.

CONTENTS

FOREWORD

As a practitioner in criminalistics for the past four decades, I am honored to write this foreword on behalf of my colleagues Barry Fisher, David Fisher, and Jason Kolowski. I have had the pleasure of working with Barry on a number of National Institute of Justice projects, and have worked alongside both Dave and Jason at the Forensic Biology Laboratory of the New York City Office of Chief Medical Examiner.

With the rapidly growing interest in forensic science, many more questions will arise from the uninitiated than can be answered by watching nonfictional television shows or from the careful review of any single Internet site dealing with criminalistics. Previous texts dealing with this subject at the student level are often difficult to comprehend. *Forensics Demystified* will provide straightforward answers to many of the questions posed by the novice for many years to come.

My colleagues, Barry Fisher, David Fisher, and Jason Kolowski, have nicely interwoven the various disciplines of forensic science into one easy to read text ranging from the crime scene, laboratory analysis, through the courtroom. Serious students of forensic science will, in particular, appreciate the clear explanation of DNA analysis, the significance of trace evidence, crime scene examinations, and forensic pathology.

Dr. Peter Pizzola
Crime Laboratory Director
New York City Police Department

PREFACE

This book is intended for people who want to become acquainted with the field of forensic science. Students, particularly high school students who are beginning to think about career possibilities, will enjoy this book as a general overview of the forensic field. For science teachers, this book reveals an applied science that is interesting for students and has real-world applications, all the while remaining firmly rooted in the scientific method. This book can be used as a guide for screenwriters, journalists, and authors, especially crime fiction novelists who want their works to contain accurate forensic science information. Attorneys and judges who have had little or no experience trying cases involving forensic evidence will welcome this book as an invaluable aid. We hope that this book can enlighten, educate, and entertain anyone who is captivated by forensic science.

Barry A. J. Fisher
bajfisher@earthlink.net

David R. Fisher
davidrfisher@hotmail.com

Jason Kolowski
Jckforn6@hotmail.com

ACKNOWLEDGMENTS

From *Barry A. J. Fisher:*

One of the pleasures of being in the forensic science profession for over 35 years is the opportunity to see the field change and to mentor a new generation of forensic scientists. Thus it is with special pleasure I thank my co-authors Jason Kolowski and David Fisher, both from the New York City Office of Chief Medical Examiner. Their enthusiasm for this project as well as their fresh approach to the subject will make this a readable book for those interested in the subject. I have no doubt that these two fine scientists will become 21st century leaders in forensic science. My boss, Sheriff Leroy D. Baca, Sheriff of Los Angeles County, deserves special mention. He and his management team encourage his staff to take proactive positions in our respective fields and I have followed that counsel to the benefit of our crime lab, the citizens of Los Angeles, and the forensic science profession. Thanks to Judy Bass and her team at McGraw-Hill for her support of us in this project. Finally, I wish to express my gratitude to my wife Susan for her patience with me. I manage to take on many significant projects and she always encourages me in my efforts.

From *David R. Fisher:*

There are many people in my life to whom I owe a tremendous amount of gratitude. Many thanks go to my parents, who encouraged me to pursue a career in the fascinating field of forensic science. Having a father as a forensic scientist to serve as a mentor is an added plus, in addition to the fact that we always had interesting dinner conversations growing up. My teachers and professors at Chatsworth High, UC San Diego, and John Jay College of Criminal Justice were instrumental in getting me hooked on science, especially Lyn Ikoma, Peter DeForest, and Tom Kubic; thank you for challenging me to think critically. I also want to thank Ravs Jeff Wohlgelernter, Shlomo Weiner, Avraham Braun, Yisroel Rokowsky, Yosef Kalatsky, and Lawrence Rothwachs, all of whom have had a profound influence on my life. Much appreciation goes to my colleagues in the Department of Forensic Biology at the New York City Office of Chief Medical Examiner. They are an extremely dedicated group of scientists, who continuously strive to use "science in the pursuit of justice." The director of the department, Dr. Mechthild Prinz, and the Chief Medical Examiner of New York City, Dr. Charles Hirsch, oversee one of the finest forensic laboratories and medical examiner's offices in the world. Much appreciation also goes to Jason Kolowski for his help in writing the book; Jason is one of those people that knows something about almost everything and was a big asset on this project. Thanks also go to Irene Wong, an extremely talented artist and forensic scientist, for illustrating the figures in the book so that the reader can better grasp the concepts being discussed. Recognition also goes to my good friend Dr. David Taylor for authoring the chapter on forensic psychiatry. Thank you to Dr. Peter Pizzola, Director of the NYPD Crime Lab, for reviewing the manuscript and offering valuable suggestions on ways to improve it. Mark Desire, JD, also reviewed the legal chapter and made helpful comments. Our editor at McGraw-Hill, Judy Bass, has tremendous amounts of patience; thank you for your help in preparing this book. Bill McManus's detailed copyediting improved the clarity of our writing. Thanks also go to Ronald D'Souza and Laura Hahn. Finally, words cannot express my appreciation and admiration for my wife and best friend Deena; without her, I would not be where I am today. Thank you for entertaining Sam and Sarah during the many hours I was trying to write. Finally, I want to express my gratitude to the Almighty for His continued blessings He bestows on us each and every day.

From *Jason Kolowski:*

Albert Einstein once said that "imagination is more important than knowledge." The people listed below are only a short list that represents the amazing network of people in my life that have fostered not only my imagination, but bestowed knowledge, wisdom, and those occasional moments of amazing clarity. While I cannot list everybody, I would like to thank the following influential people in no particular order: H. David Kay, Peter De Forest, Anna Duggar, Margaret Wallace, Mechthild Prinz, Paul Goncharoff, Robert Shaler, Jaime Negron, Jeff Cardon, Craig Tepper, and Tony Watson. I would also like to thank my parents for not asking too many questions when I decided to become a scientist and not a teacher, and to my extended family—thanks for all the love and support. A huge thank you is due to Irene Wong for contributing her talents to this work. I have the deepest admiration for Barry Fisher and Dave Fisher for bringing me into this project, and offer a heartfelt thanks to Judy Bass and her team of people at McGraw-Hill for making this book a reality. Also, a big thank you to Ronald D'Souza and his team at TechBooks International for all of their hard work. To Pete Pizzola, the former director of the Yonkers Crime Lab, former member of the OCME-FARU team, and current director of the NYPD crime lab—thank you buddy, your input and suggestions were much appreciated. Finally, they say that behind every great know-it-all there is a woman rolling her eyes. That woman is my lovely and talented wife, Stacy. She has challenged me, completed me, and helped me achieve my full potential. Thank you!

Justice, only justice shalt thou pursue...
– Deuteronomy 16:20

CHAPTER 1

Introduction to Forensic Science

Forensic science is a strange profession. If you meet a forensic scientist and ask what it is that they do, you might get a variety of answers, which could include anything from "I examine bullets collected from crime scenes" to "I sequence mitochondrial DNA." In the field of forensics, almost every professional discipline is represented—there are doctors, lawyers, engineers, dentists, geneticists, nurses, botanists, and computer programmers. All of these professionals work in their specialty as forensic examiners, and they all work toward the ultimate goal in forensic science: to uncover the facts of an investigation by using the scientific method.

According to *Merriam Webster's Collegiate Dictionary*, the word *forensic* is defined as "belonging to, used in, or suitable to courts of judicature or to public discussion and debate." *Forensic science* can be fully thought of as the scientific investigation into matters pertaining to law in the course of a crime. An easier way to think about forensics is: the practice of matching people to people, or people to places, during the investigation of a crime. The reason, therefore, that so many different people and professions are involved in forensics is that crimes happen all the time in different settings and in different ways. When the boss of an international company is caught stealing money, a forensic medical doctor would not be of much use in the investigation, but a forensic accountant would be. Forensics is also not limited to just criminal investigations; the science of an investigation can be applied to civil cases as well. When a set of brakes on a car fails and all the passengers are killed in an accident, a forensic fingerprint examiner would not contribute much to the investigation, but a forensic engineer would be vital in providing evidence for a civil case.

When people think about forensic science, they usually picture the technicians at a crime scene, dusting for fingerprints and photographing blood spatter, or searching through a room with an alternate light source to find trace hairs and fibers left by the suspect. This type of forensic science is known as *criminalistics*. Forensic scientists who work in this field refer to themselves as *criminalists*, or ones who investigate crimes.

Criminologist vs. Criminalist

Careful! Don't confuse a "criminalist" with a "criminologist." A criminologist is a person who studies criminals and their behavior from a social and psychological aspect. A forensic psychologist who studies criminal profiling is an example of a criminologist, *not* a criminalist.

Criminalists typically are educated in forensic science programs and tend to be well versed in all aspects of forensics, but they can specialize as well. The subsections of *forensic DNA analysis* and *ballistics*, for example, are sometimes grouped under the "criminalistics" heading, but people working in these fields are specialists in their own right. The people that everybody associates with forensic science, however, are truly the specialists in the criminalistics field. These people include fingerprint examiners, trace evidence analysts, hair and fiber specialists, microscopists, and crime scene examiners. The forensic scientists working in these fields need a strong foundation in the natural sciences, background in all aspects of forensics, as well as a large amount of hands-on experience in their specializations.

Typically, a criminalist has an undergraduate and/or graduate degree in forensic science from an accredited university, and has put in additional time through advanced courses or on-the-job training to become an expert in a particular field. For forensic specialists outside the criminalistics field, the opposite is true. For example, a forensic pathologist (a medical doctor who performs autopsies) first attends medical school and becomes a medical doctor before specializing in forensic pathology. The same is true for forensic accountants, lawyers, and engineers as well.

The Forensic Science Laboratory

Because forensics can have such a diverse number of people working in the field, no two forensic laboratories are the same. Forensic labs are commonly built to serve a specific population with the services that are needed most—whether it is at the local, county, state, or federal level. For example, a forensic DNA lab is a fairly complex and expensive lab to maintain for a small town, and therefore these labs are typically located only in large cities, or at the state or federal level. On the other hand, a small town might have a huge drug problem, and thus it's fairly common for a local forensic science lab to have drug analysis capabilities, as well as fingerprinting and photography capabilities.

In an ideal setting, every major city would have a laboratory building dedicated to forensic investigations, which would house the following major subsections of forensic science:

1. hair and fiber/trace analysis
2. drugs and toxicology
3. forensic photography/digital evidence
4. ballistics and firearms analysis
5. dental/odontology
6. anthropology
7. fingerprinting
8. medical examiner/coroner
9. questioned documents
10. serology/forensic biology (DNA)
11. evidence collection teams (CSI)
12. evidence intake and security
13. records and archive section
14. support staff

Each one of the listed subsections requires a separate and independent workspace/lab, although some subsections can be combined together, such as hair and fiber and serology, since some of the same equipment is used in each.

One of the most important aspects of a forensic lab is the way in which evidence is brought in, catalogued, and stored before, during, and after analysis. This job is typically handled by uniformed police officers or evidence control specialists. Because it is absolutely critical to account for the *chain of custody* (the documented record of the existence and handling of a piece of evidence) in an investigation, these people are on the front lines of maintaining the integrity and security of evidence when it enters the lab. Usually the evidence control unit is a centralized, single unit for a lab, and the different departments come to that unit to request the evidence for processing.

Because the chain of custody is presentable in court, the evidence must be accounted for during lab analysis by being in the possession of a person, and must always be stored in a secure location within the lab. When the analysis is complete, all of the evidence must be retained in the lab or sent back to the evidence control unit. Evidence is rarely discarded, because it must be made available for further testing, if necessary. Further testing might be performed in the same lab to help confirm the previous results, or might be sent to another lab at the request of a defense attorney to allow for a second opinion. If the evidence is ever discarded, misplaced, or lost, then all of the work done in the case could be in jeopardy and the case could be thrown out of court. There are hundreds of examples where sloppy, inconsistent, or missing chains of custody were enough to have cases thrown out by a judge.

Regardless of the size of the lab or the services that it provides, a good forensic science lab must always be ready to respond to the needs of the community that it serves. In certain geographical areas, the forensic investigation services are part of the police force; in others, the police or law-enforcement sections are separate from the forensic science lab. As long as there is a professional-understanding and open communication between the police and the forensic lab, the goals of each can be met and the community will benefit in the end. This also extends from one forensic lab to other forensic labs. In smaller population areas, where DNA services are not locally available, the local forensic lab requires the services of a larger forensic lab to process DNA samples, and this might be at the state or federal level. The same might be true for other services like questioned documents or hazardous material testing, which might not be locally available. Again, by having open lines of communication and professional contacts within the field, all of the forensic labs can benefit from the services that are available throughout the state, country, or even the world.

Modern Forensic Science

The field of forensic science has numerous different aspects. Regardless of which field or specialty a forensic scientist practices, the way the forensic investigation is done is always the same. Forensic science uses the *scientific method* to carry out investigations. The scientific method is a simple, structured way of problem solving, and can be thought of as a circular system. First, the scientist identifies the problem or question. Next, the scientist forms a *hypothesis* (an educated assumption or guess) about the solution to the problem or question. Using the hypothesis, the scientist performs experiments to attempt to disprove the hypothesis. If the hypothesis can be disproved by experimental data, then the scientist must modify or rethink the hypothesis, and begin the experiments again based on the revised hypothesis. This is known as the *feedback loop*, and it is what makes the process circular. If the experiments fail to disprove the hypothesis, the scientist can accept the hypothesis as the answer to the original problem, with a reasonable degree of scientific certainty.

Although the scientific method sounds complex, it can be done on even relatively simple matters. For example, if you were to lose your keys, you would first identify the problem (I lost my keys!). Then, your hypothesis to solve the mystery of the lost keys might be, "The last time I knew I had my keys, I was at my desk, so the keys must be on my desk." To test your hypothesis, you perform the "experiment" of examining the desk area. You do not find your keys there, so your experiment has disproved your hypothesis. You then must modify your hypothesis. You remember that on your way out the door, you had your keys in hand, but were stopped by a co-worker struggling with a large box. Your revised hypothesis, then, is, "I put my keys on the co-worker's desk to lend a hand with the box." To test this theory, you search the co-worker's desk. You find your keys there. Since there is no way to disprove your new hypothesis (in fact, you've proven it beyond the shadow of a doubt), the hypothesis is valid and the method is complete.

In forensic science, the scientific method is applied rigorously and consistently by all forensic examiners. For example, let's say a used condom arrives at a lab as the only evidence of a rape, along with DNA samples from the female victim and male suspect in the case (Chapter 7 covers DNA in depth). The hypothesis in this situation is that the suspect raped the victim while wearing the condom, and therefore DNA from both the suspect and the victim should be present on the condom. By performing DNA testing, the DNA from the victim is matched to DNA on the condom, but the suspect's DNA is proven to be different from the male DNA on the condom. Therefore, the suspect in this case does not match the DNA evidence, and the hypothesis is disproved. The hypothesis can then be modified to answer the question, "If it is not the suspect's male DNA on the condom, then whose DNA is it?" A quick database search could come up with the answer, which would disprove all other hypotheses. Additional hypotheses may have to be formed to consider other possibilities.

Identification vs. Individualization

By utilizing the scientific method, forensic science is a powerful tool that helps solve crimes. By applying the scientific method to each and every case, forensic scientists approach each and every question that one could encounter during an investigation with scientific rigor. However, regardless of the crime and regardless of the evidence, there are only two issues that need to be resolved by a forensic scientist: identification and individualization.

In forensic science, *identification* is simply answering the question "what is it?" However, sometimes this is not as easy as it seems. Imagine being given a piece of tape with several small grains of debris stuck to the adhesive. You are asked to identify the material. Is it sand? Salt? Dirt? Drugs? To answer this question, you must use the scientific method by collecting some information about the debris. So, suppose you take a look at the material under a microscope. Under magnification, the grains appear to be opaque white cubes, and you hypothesize that it is salt. Thus, you perform a simple chemical test on a grain or two for the presence of sodium and chloride (sodium chloride = basic table salt). The test is positive, meaning you've identified salt! (And, you have performed the job of a trace evidence examiner.)

Now imagine that you are given that same piece of tape with the salt stuck to it and asked "what brand of salt is it?" That question is a lot harder to answer, because you have to individualize the salt. *Individualization* can be thought of as the uniqueness of an object to the exclusion of all other objects like itself. Individualizing salt is tough, because all salt is formed from the same basic elements, sodium and chloride. What sets one type of salt apart from another are the impurities, or the things in the salt that are not salt—other minerals, trace elements, even the amount of water can differentiate one type of salt from another. But, to find these trace aspects of the salt, even more sensitive or detailed tests might be necessary.

As an example, let's say you are a forensic drug analyst. One day, a detective comes to your lab with an unknown white powder and wants to know what it is. There are a number of different ways to test the unknown, but a simple and decisive test is a *microcrystal test*. Such a test is done on a microscope slide using a small amount of a powder that reacts with different chemicals to form unique crystals. The crystals that form on your slide indicate that the bag is filled with cocaine, which you tell the detective. Then, the detective wants to know if it's the same cocaine from a drug bust from a month ago. Because this now requires you to individualize the cocaine apart from all other batches of cocaine, a more specific and sensitive test is needed, possibly using a highly specialized instrument called a gas chromatograph-mass spectrometer (GC-MS) to identify the trace elements and adulterants that make this cocaine unique. When you run this test, you find that it is different from the batch of cocaine from the previous month, and in fact has never been seen before in your lab. Based on the tests, then, the identification and individualization of the unknown is complete.

Another important concept regarding identification and individualization is the distinction between macroscopic traits and microscopic traits. In the example of salt grains on the tape lift, it was necessary to look at the microscopic traits to make the identification and then go to finer details to individualize the salt. Sometimes, though, the identification is easily done on the macroscopic level, and the individualization is left to the microscopic level.

For example, take five pennies and lay them on a table. It's easy enough to identify them by simply looking at them—they are pennies. But stand back 5 to 6 feet from the table. Can you tell the difference between each one of the pennies? Probably not (although you might be able to if some are darker from being tarnished). As you move in closer and closer, you start to distinguish more and more differences between them. At about 1 foot from the pennies, can you tell them apart? You should be able to notice some differences. The pennies might have different dates on them, or different symbols indicating the mint where they were created. Even though you are at a macro level, you can begin to individualize the pennies better than you can by standing 5 feet away. Now suppose that all of the pennies are from the same mint and were produced the same year. At 1 foot away, you would not be able to tell them apart. But with a magnifying glass, you could begin to notice the fine details on each individual penny. Once again, you move in closer to a micro level and begin to individualize the pennies. By beginning with the macroscopic traits and working down to the microscopic traits (or even smaller!), forensic scientists can begin answering the questions of identification and individualization.

Evidence and Exemplars

In the course of a forensic investigation, a forensic scientist examines items from a case to begin to answer the questions of identification and individualization. These items are known as *evidence*. Most people are familiar with evidence because of police shows on television and the movies, but they have probably never thought much about it. In forensic science, evidence is used as physical proof. Without evidence, there can be no proof, and therefore no justice.

In forensic science, evidence usually takes the form of an *unknown* in the lab. It is the forensic scientist's responsibility to test this unknown and determine what it is (identification) and what sets it apart from other evidence (individualization). To do this, forensics utilizes *exemplars*, or "knowns." Exemplars are objects that are commonly encountered in investigations or the possible real object that created the evidence. For example, if a tire tread is found at a crime scene, it becomes evidence. The tread mark will be photographed, measured, and molded for analysis. Examination of this tire tread (evidence) can tell the forensic scientist the make and model of the tire or car, the size or weight of the car, and the speed and direction that the car was moving. If a suspected car is uncovered in the investigation, the tires on the car can serve as exemplars, and each will be tested and measured to see if it could have produced the tire tread from the crime scene. In such a case, because of wear and tear on the tire, it might be possible to individualize the tire back to the exact tire on the car, but only if the known tire can be compared alongside the evidence. ✳

Every facet of forensic science relies on knowns and unknowns for the process of forensic testing. In a drug or toxicology lab, known chemical signatures of drugs and toxins must be obtained to compare against the evidence that is tested in the lab. In a firearms lab, an exemplar pistol (e.g. found on a suspect) is able to re-create the markings on a bullet to see if it matches a bullet from the evidence in a shooting. In a DNA lab, the DNA profile of a suspect must be obtained if it is to be compared against the blood or semen evidence from a violent crime. In a questioned documents lab, the handwriting analyst must have exemplar handwriting samples from a suspect to compare against a handwritten ransom note as evidence. Only by comparing knowns against unknowns and vice versa can a forensic scientist truly answer the questions of identification and individualization.

Inclusions and Exclusions

When the evidence in a case is linked to an exemplar or known, it is called an *inclusion*. Inclusions typically occur when the evidence is individualized to a known, apart from all other knowns in the case. For example, if the broken end of a lock pick that is recovered from the scene of a burglary matches perfectly to the pick recovered on a suspect, there is an inclusion of the evidence (the broken piece of the lock pick) to the exemplar (the pick found on the suspect). However, if the evidence does not match the exemplar in a case, then there is an *exclusion*, and that exemplar is ruled out from having produced the evidence. Similarly, suppose a suspect in an arson investigation is arrested and accelerant is found in the trunk of his car; it can be tested against the accelerant used in the fire. If the flammable liquid from the trunk proves to be chemically different from what was used to start the fire, that exemplar accelerant is excluded from the investigation.

Inclusions and exclusions are also both equally important in forensics. Inclusions not only aid in matching people to people and people to places in the course of the investigation, but also, by establishing these links, open more avenues or clues for the investigation. Likewise, exclusions are important because they help eliminate the suspects, places, or possible events that surround a crime and can aid in narrowing down the focus of the investigation.

Sometimes, there are instances in forensics where the evidence is presented in such a way that it excludes all other possible exemplars. This is commonly seen in DNA testing, where despite a "match" of an unknown DNA profile to a known DNA profile, the statistical relevance of the match is what matters. For example, the chance of someone drawn at random from the population having the same DNA profile as the suspect would be greater than one in a trillion. The inference is that the known profile is so rare, and since the suspect's DNA profile is the same as the evidence profile, there is little room for doubt that the donor of the DNA is the contributor of the evidence. This can be tricky, and will be discussed further in Chapter 7, which covers DNA analysis. The main thing to keep in mind with forensics is that inclusions and exclusions are the keys to uncovering the truth in any investigation. By applying the scientific method to the evidence and exemplars of the case, inclusions and exclusions will help answer the questions of identification and individualization, and aid the forensic scientist in solving the mystery of the case.

The History of Forensic Science

Because forensic science is such a diverse field, it's always growing and changing as new professions arise. (In the 1890s, around the time forensic science was just starting out, nobody could have known that a forensic computer analyst would ever be necessary to catch a white-collar criminal.) Forensic science as a profession actually got its start in Europe in the mid-1800s. The Industrial Revolution brought more and more people into cities, and as the cities grew, so did the crime rate. But, the change that brought about huge increases in the production and distribution of goods and services also led to increases in science and technology. Forensic science was born into this age of reason and industry, but it progressed slowly. As it gained in popularity and reputation, forensic science began to provide evidence instead of *hearsay* (testimony from a person who did not directly witness a crime) and false eye-witness testimony. As the burden of proof shifted from the accused (trying to defend their innocence) to the prosecution, both the courts and the public became more and more aware of the exciting new field that was emerging and helping to solve crimes. Several key people throughout this time were instrumental in the development of forensic science into the field that it is today.

GENERAL CRIMINALISTICS

The actual term "criminalistics" is credited to Hans Gross, an Austrian, who published one of the first books on the subject in 1893. Gross's book was the first of its kind to show the uses of microscopy, chemistry, botany, and physics in forensic science along with the classical backgrounds of serology, ballistics, and toxicology. Gross's foundations and suggestions laid the early framework for generations of forensic scientists, and began the long journey of bridging the philosophy of forensic principles with science.

Also in Europe, in the early 1900s, Edmund Locard founded the first forensic science laboratory in Lyon, France, and his work became one of the foundations of modern-day criminalistics. Locard believed that every time two objects come into contact, there is an equal and consistent exchange that takes place between the two objects. This has become known as Locard's Exchange Principle (see the corresponding sidebar), and has been validated time and time again in all aspects of criminalistics. Locard also developed the basic rules of fingerprint analysis, and his work on all aspects of criminalistics shaped modern forensic science.

As the field of criminalistics began to grow in Europe in the early 20th century, forensic science was emerging in the United States at the same time. In the 1920s, Calvin Goddard and August Vollmer established the first forensic science laboratories in the United States. Vollmer was the chief of police in Berkeley, California, and is

Locard's Exchange Principal

Locard's Exchange Principle states that "with contact between two items, there will be an exchange." For example, if a person touches the surface of a painted wall with their fingers, the wall and fingers interact and exchange trace evidence. The fingers pick up dust, flakes of paint, even trace metals on the wall, and the wall likewise picks up oils and cells from the finger, as well as any trace evidence that was on the fingers from before the contact. Transfer can occur at a cellular level (like the finger on the wall) all the way up to a macro level, as seen by paint marks on guide rails along the highways from errant vehicles—the cars that left their paint marks on these rails no doubt picked up some of the metal from the rails as well. Each of these is a clear example of Locard's Exchange Principle at work, and underscores one of the most important aspects in all of forensics: one must "listen" to the evidence. The forensic scientist's job is to uncover and reconstruct how the evidence fits into the investigation of a crime.

credited with the creation in 1924 of the first forensic science lab in the United States. Goddard was a scientist in Chicago, Illinois, and his investigation of the St. Valentine's Day Massacre of 1929 showed the strength of forensic science and led to the creation of a major laboratory at Northwestern University.

Vollmer's lab began as a small room behind his office, and grew into one of the most well-established forensic science labs in the world. As the Berkeley lab expanded, Vollmer enlisted the help of two notable people, Edgar O. Heinrich and Paul Kirk. Heinrich was a meticulous scientist and an expert in trace analysis. In one famous investigation (see the sidebar "The Wizard of Berkeley"), after only examining a pair of coveralls recovered from the scene of a train robbery, Heinrich was able to provide to the police not only a physical description of the suspect, but also clear details of the suspect's habits and personality as well.

Paul Kirk is considered by many to be the father of modern American forensics. After working for years in the Berkeley lab, he and August Vollmer founded one of the first schools for forensic science at the University of California at Berkeley, and Kirk himself taught and oversaw the criminalistics section of the program. He also

The Wizard of Berkeley

In 1923, a train car was robbed by three men. The only item recovered from the scene of the crime was a pair of coveralls. These were sent to the Berkeley lab and analyzed by Edgar O. Heinrich. Heinrich, an expert in microscopy and trace evidence, carefully compiled all of the minute details from these coveralls, and within a week or two, issued his findings to the police.

Heinrich reported that the suspect in the case (the wearer of the coveralls) was a white male with a light complexion, medium-brown hair, light-brown eyebrows, a mustache, and a height of around 5'10". What was strange about the results was that Heinrich also reported that the suspect was left handed, was most likely a logger or woodsman in the Pacific Northwest, was well groomed to the degree of being meticulous about his appearance, and, when caught, would be smoking and wearing a jacket and bowler hat!

As strange as Heinrich's analysis was, he was right. When the police arrested one of the d'Autremont brothers, the man confessed to everything about the train robbery, all the while sitting straight up in his new coat and bowler hat, smoking a cigarette! Heinrich's analysis of the coveralls had uncovered many aspects of the suspect's appearance, such as head and eyebrow hairs for the hair color, and the length of the coveralls corresponded to the height of the suspect. The victims of the robbery had seen the robber's hands, which Heinrich included in the report about the skin color.

What had prompted the rest of Heinrich's conclusions was the analysis of the contents of the coverall's pockets. Heinrich had found wood chips and pine needles in the right-hand pocket of the coveralls. He was able to identify the species of the tree as that of a fir tree specific to the Pacific Northwest. He concluded based on the wood chips that the suspect was a woodcutter or logger, and that he was left handed, because by gripping an ax with his stronger left hand, the suspect's right-hand pocket would be turned toward the tree and collect wood chips as they flew around. Heinrich also found tobacco, wax, clipped fingernails, and the burnt ends of rolled cigarette butts in the upper-right breast pocket of the coveralls. Most impressively, based on the wax, which he concluded was moustache wax, he surmised that the suspect might be very vain about his appearance, because it would be odd for a person involved in a job like logging to use moustache wax. Also, Heinrich concluded, by keeping the fingernail clippings and cigarette butts rather than tossing them on the ground, the suspect was extremely meticulous about his appearance and actions. Because of this personality type, Heinrich concluded that that the suspect, having just robbed a train, would be dressed in the most popular style of the day, which was that of the coat and bowler hat. Sure enough, Heinrich was right, and although no forensic scientist today would make such outlandish claims based on the evidence, it is a unique and amazing example of the power of criminalistics as applied by the wizard of Berkeley, Edgar O. Heinrich.

published one of the first forensic science textbooks, *Crime Investigation*, to provide students and practitioners alike not only the fundamentals of forensics, but the theories and principles behind the science as well. For several years, the UC Berkeley program educated some of the most influential forensic scientists in the country, and promoted unprecedented advancements in the science and technology used in forensic investigations. Several graduates of Kirk's program went on to establish other forensic science programs, such as Peter R. De Forest at the City University of New York's John Jay College of Criminal Justice, and George Sensabaugh, who took over the UC Berkeley program. The influence that Kirk has had over the foundation of modern forensics is recognized in the Paul Kirk Award, which is awarded by the American Academy of Forensic Sciences every year to an individual in the forensic community who has fostered outstanding contributions to the advancement of forensic science.

OTHER FORENSIC DISCIPLINES

As general criminalistics has expanded and developed over the years, so too have the other forensic disciplines. With each technological revolution and scientific advancement, there has been a similar increase in the knowledge and techniques in such forensic fields as toxicology, fingerprinting, questioned documents, DNA analysis, and many others. To understand the impact of the newest developments, it is important to go back and look at the origins of each of these fields.

Toxicology

For several hundred years leading up to the 1900s, arsenic was known as "inheritance powder." This powerful poison was used to kill off kings and queens, wealthy heirs, and troublesome relatives. The reason that arsenic was used so successfully was that the person who was poisoned simply passed out and died, apparently from natural causes. This all changed thanks to the work of two men, Mathieu Orfila and James Marsh. Orfila is considered the father of modern *toxicology* (the science and analysis of toxins and poisons). Orfila was a Spaniard who in the 1820s became the leading professor of medical chemistry at the University of Paris. His early writings and research led to the development of the field of forensic toxicology, where James Marsh, a Scotsman, later developed the Marsh Arsenic Test in the 1830s. Using Marsh's test, it could be proven that arsenic was used to commit murder by testing the blood or remains of the victim. The test is even sensitive enough to find trace amounts of arsenic in food and liquid residue from plates and cups that the victim used in their "final" meal. The Marsh Arsenic Test is still used in several toxicology labs around the world as a simple and rapid test for the presence of arsenic.

Fingerprinting and Pattern Recognition

Take a minute or two and examine the ends of your own fingers. Any one of your fingerprints, if left at a crime scene, can be used to individualize you apart from all other people in the world. However, before fingerprinting became widely known and accepted as a tool for identification, another system, called *Bertillonage*, was used worldwide. Bertillonage, named for its discoverer, Alphonse Bertillon, was a series of measurements of various body dimensions that could be used to "individualize" a person. Certain measurements, such as the width of the eyes or the length of the hand, were recorded on a card along with a photo of an individual. In theory, Bertillonage would be an ideal system for identifying a person, except that two major problems were encountered.

First, there was no standardization in the taking of measurements; for example, when measuring the distance between eyes, was the measurement taken from pupil to pupil, or outer edge of the eye to the other outer edge? The second problem was there was no firm proof that the Bertillonage system was actually capable of individualizing one person from another. Because of these issues, and a famous case of mistaken identity (see the sidebar "The Will West Case"), Bertillonage was abandoned and fingerprinting came to be the most reliable system of individualization throughout the 19th and most of the 20th century.

The science of fingerprinting became known to forensic scientists due to the work done by Francis Galton, Henry Faulds, and Edward Henry in the late 19th and early 20th centuries. Galton and Faulds were among the first scientists to recognize the importance of, and to use, fingerprints for forensic identification and individu-

The Will West Case

In 1903, a black man named Will West was arrested and sent to Leavenworth Federal Penitentiary in Kansas. When his Bertillonage measurements were taken and filed away, the prison officials were surprised to find another prisoner already incarcerated with the same name and Bertillonage measurements. The second man, named William West, was already serving a sentence in Leavenworth. When the photographs and Bertillonage measurements of the two men were compared, they were indistinguishable. Side by side, the two men could almost pass for identical twins, and there is some rumor to this day that they might have been separated at birth. In the end, it was shown that the only true way to individualize the two men was by fingerprint comparison, and the Will West case quickly became famous for proving the strength of fingerprint comparison and marked the beginning of the end of Bertillonage.

alization purposes. Edward Henry was the head of Scotland Yard in London at the turn of the 20th century and developed a classification system for "cataloging" fingerprints based on common patterns that are present on each finger. This became known as the Henry Classification System. Today, the Henry Classification System of fingerprinting has been replaced by the computerized worldwide database of the Automated Fingerprint Identification System (AFIS).

Questioned Documents

For several thousand years, humans have used written language to communicate, to record, and to entertain. Handwriting, from quickly jotted notes to elegantly formed calligraphy, has the ability to capture the nuances from the person with the pen (or quill, pencil, brush, etc.) in hand. These nuances can be shown to be consistently formed over time by an individual, and therefore aid in the forensic investigation of a crime. However, it wasn't until 1910 that Albert Osborn recognized the individualizing nature of handwriting and various other questioned document techniques, and became one of the leading figures in the field of questioned documents. Many of the techniques and rules of handwriting analysis that Osborn developed are still in use today. However, advancements in technology have also forced questioned document examiners to look beyond handwriting to examine typewriters, Xerox machines, printers, and fax machines, as well as the paper and inks that are used in the creation of such documents. One example of such a modern investigation was done on the infamous "Hitler Diaries." In 1983, a German magazine claimed to possess the actual diaries of Adolf Hitler. When these diaries were examined by questioned document experts, not only did they find classic signs of forgery based on Osborn's rules, but the tests on the paper and ink proved that both were produced decades after Hitler's death!

As developments in printing and copying have progressed throughout the years, so have the techniques of questioned document examiners. With high-tech tools such as infrared photography and the electrostatic detection apparatus (ESDA), combined with Osborn's classic rules, forensic questioned document examination is able to keep up with demands of a technologically driven society.

Serology and DNA Analysis

Blood and body fluids have been used since the early 1900s to help forensic scientists establish the circumstances and people surrounding a crime. When Karl Landsteiner discovered the ABO blood-typing system in 1900 for the medical community, it was quickly seized upon by the forensic scientists of the day, and used to help include or eliminate potential suspects of a crime. Likewise, the similar discovery of the Rhesus factor by Alexander Wiener in 1940 (at the New York City Office of Chief Medical Examiner), and the various other blood subgroups discovered later on, was also utilized by the forensic community for identification purposes. Throughout the years, as medical knowledge of blood and other body fluids increased, so did the forensic knowledge of how these blood groups could be applied in a criminal justice setting.

Whereas the discoveries of the ABO blood-typing system and the Rhesus factor led to rapid development of forensic techniques, the discovery of DNA did not have a direct effect on forensic science for nearly forty years. In 1953, James Watson and

Francis Crick, using x-ray crystallography techniques developed by Rosalind Franklin, uncovered the structure and function of DNA (deoxyribonucleic acid). Although much research and work was done over the years from an academic aspect, there was only a limited application of forensic DNA use in the late 1970s and early 1980s. However, it was clear to the forensic community from the beginning that DNA was a truly remarkable means of individualization and identification. Based on the rules of genetic inheritance, no two people should have the same DNA, and, therefore, it was a unique feature! During this time, however, forensic work on DNA for identification and individualization was limited because the tests required large amounts of blood or body fluids for analysis, and the results were not always reliable due to the lack of samples and the novelty of the science.

In the 1980s, one of the first true forensic DNA tests was created by Sir Alec Jeffreys, which came to be known (although not accurately) as "genetic fingerprinting." Jeffreys developed a basic system of digesting DNA with enzymes that caused the DNA to break down into specific-sized fragments that are unique to each person. These fragments of DNA were then separated based on their size using a technique called *electrophoresis* (the process of separating substances based on size and charge using a permeable medium and an electrical charge; see Chapter 7 for more information), and what developed was a "fingerprint" of the DNA in the electrophoresis gel image. Jeffreys became famous for this technique when it was used not only to find the murderer of two English girls, but also to exonerate an innocent suspect in the same case (see the sidebar "The Colin Pitchfork Case").

The Colin Pitchfork Case

In 1983, the body of a 15-year-old girl was found raped and murdered in the fields outside the English town of Narborough. A semen sample was collected from the body and tested using the traditional methods of the day, which were *secretor status* (a "secretor" has the ABO blood-type antibodies in other body fluids, such as semen and saliva) and *enzyme typing* (a forensic test that screens for the presence and structure of proteins and enzymes to look for unique factors that can be used to identify/individualize a person). The secretor status was positive and the enzyme test showed a rare enzyme type, but without a suspect to compare against, nothing more could be done.

Then, in 1986, the body of another 15-year-old girl was found in the same place as before, and again the serological tests gave the same result. This time, though, the police had a suspect in custody. Richard Buckland, a local man who found the second body, was questioned by the police and admitted to the recent murder but denied involvement with the previous one. Serological tests on Buckland's blood showed he was of the same blood and enzyme type as the unknown attacker, and he was placed under arrest. Buckland's family, believing him to be innocent, contacted Alec Jeffreys to come and perform his new DNA tests on the samples in the case. When Jeffreys did so, the DNA tests proved that the two murders were committed by the same man, but that man was not Richard Buckland. With those DNA tests, Buckland became the first wrongly accused person to be freed based on DNA testing, but the real killer was still at large.

Since the police now believed that Jeffreys' tests could bring the villain to justice, they began a systematic testing of all males over a certain age in the town of Narborough. Over 5000 individuals were tested using Jeffreys' "genetic fingerprinting," but none was found to match the semen taken from the two girls. Then, as luck would have it, a local man named Ian Kelly was overheard bragging about how he had masqueraded as his friend Colin Pitchfork when the police were taking blood samples from the men in the town. When this was reported to the police, the police tracked down Pitchfork and finally found the match to the DNA samples that had eluded them for so long. Colin Pitchfork was convicted in 1988 for the rape and murder of the two young girls, and is still in prison today as the first person to have ever been found guilty based on forensic DNA testing.

Around the same time as the Colin Pitchfork case was being investigated in England by Jeffreys, the next great evolution in DNA analysis was occurring in America. In 1983, while driving on a windy stretch of California highway, an American scientist (and surfer) named Kerry Mullis realized that DNA could be copied thousands and millions of times *in vitro* (in a test tube) by mimicking natural biochemical processes that occurred inside living cells. His discovery became known as the *polymerase chain reaction*, or *PCR* for short, and virtually overnight changed the way DNA could be used in forensic science. Because PCR could copy DNA in a lab setup, it became possible to produce high levels of testable DNA from very small biological samples. The tests could also be run in such a way as to confirm the results, lending accuracy to forensic DNA analysis. Today, DNA is quite possibly the single most used tool in a forensic scientist's arsenal for true and unambiguous individualization.

Quiz

Refer to the text in this chapter if necessary. Answers are located in the back of the book.

1. Who is credited with the formation of the first forensic science laboratory in the United States?

    (a) Edmund Locard

    (b) Jacob Gilliay

    (c) August Vollmer

    (d) Paul Kirk

2. The measurement of physical features for individualization is known as:

    (a) Profiling

    (b) Bertillonage

    (c) Sociology

    (d) Anthropology

3. Which of the following items can be individualized?

    (a) Air

    (b) Undyed cotton fibers

    (c) Pennies

    (d) Dust

4. Mathieu Orfila is famous for his contributions to the field of:

    (a) Forensic science

    (b) Histology

    (c) Dactylography

    (d) Toxicology

5. Bertillonage was replaced by latent print examination due to what reason?

    (a) The discovery of fingerprints by August Vollmer

    (b) Inconsistent measurement techniques

    (c) The misidentification of two men with the same name

    (d) The death of Bertillon

6. The Marsh Test can prove the presence of:

   (a) Heroin

   (b) Arsenic

   (c) Cocaine

   (d) Copper

7. AFIS is an acronym for:

   (a) Automated Forensic Identification System

   (b) Automated Fingerprint Identification System

   (c) Automated Fingerprint Individualization System

   (d) Automatic Forensic Individualization System

8. Individualization of a person can best be achieved by:

   (a) Fingerprinting

   (b) Hair analysis

   (c) Both A and B

   (d) None of the above

9. The Henry Classification System of latent fingerprint analysis is named for which individual?

   (a) Henry Faulds

   (b) William Henry Herschel

   (c) Edward Henry

   (d) Henry Battley

10. X-ray crystallography techniques that were used to uncover the structure of DNA were originally developed by:

   (a) James Watson

   (b) Rosalind Franklin

   (c) Kerry Mullis

   (d) Francis Crick

CHAPTER 2

Legal

One of the most essential parts of being a forensic scientist is testifying about one's findings in a court of law. This is where all of the investigating, examining, gathering, interpreting, and report writing come together. The forensic scientist is called upon as an expert witness to communicate and interpret the physical evidence in a case. Experts are obligated to present their conclusions, in addition to the methods used to generate them, accurately and truthfully. This must be done in layman's terms so that the men and women of a jury and/or the judge can understand. It is critical that this testimony be based solely on the facts of the case and be rooted in sound and reliable science, because often a defendant's life and liberty are at stake.

The Expert Witness

What differentiates an expert witness from other witnesses? While a universally agreed-upon definition of "expert" does not exist, the generally accepted standard is that an expert possesses knowledge and/or skills that the average person lacks. The expert's testimony, therefore, assists the *trier of fact* (the judge or jury) in resolving the questions at hand in a court case presumably because the expert has more knowledge and experience than the members of the jury or the court. Consequently, to aid the trier of fact in the search for the truth, the expert is permitted to render an opinion during testimony and draw conclusions. Nonexpert witnesses, alternatively, may only testify to what they directly saw or know firsthand and generally may not offer their opinion about what took place. Testimony by a witness that relates to events about which the witness does not have firsthand knowledge is classified as *hearsay* and usually is inadmissible. For example, if an eyewitness to an accident later tells another person what she saw, the second person's testimony about the original incident is hearsay. The primary reason for this rule is that the opposing party has no ability to confront the person who has firsthand knowledge of the event.

When rendering an opinion, it is incumbent upon the expert to testify only within one's area of expertise. Sometimes experts testify outside of their areas of expertise simply because they have been deemed an "expert" by a court that may not fully understand the boundaries of the expert's discipline. It is for this reason that forensic scientists should have utmost integrity and be of good moral character. If an expert is asked something outside of their area of expertise, it is their responsibility to alert the court that they cannot testify in this area. For example, if a DNA expert were asked a question about a microscopic hair examination done on a head hair, it would be improper for the witness to answer, if in fact they lacked the necessary training and experience required of a trace evidence analyst conducting hair comparisons.

One of the major difficulties a forensic expert witness faces when testifying is the fact that they are required to testify in an adversarial legal system. The very nature of a trial forces the forensic expert into an adversarial role, instead of the objective presenter of scientific truth. Instead of objectively presenting the truth, which scientists mainly try to do, the forensic expert appears in a courtroom where our justice system is mainly one that pursues relative justice, instead of absolute truth. To demonstrate this reality, let us consider a yes or no question posed by an attorney to a forensic expert in the witness box. The attorney only desires a "yes" or "no" answer, yet the expert knows that a one-word answer without explanation may not convey the full meaning of what the expert wants to present. This forces the forensic scientist into a difficult position, as a witness cannot testify to things that they are not asked. This, however, is the nature of the beast, and as discussed in Chapter 1, forensic science is the application of the sciences to the law!

Voir Dire

What qualifies someone to be an expert? This question is usually addressed at the beginning of an expert's testimony. The attorney that calls the expert as their witness will ask the expert a series of questions about their qualifications. These questions focus on the expert's education, experience, publications, certifications, and membership in professional organizations. Opposing counsel will then have the opportunity to further ask the expert about their qualifications and can discredit the witness from being qualified as an expert. This process is known as *voir dire* (pronounced vwär-dir), and literally means, "to speak the truth."

It is critical that expert witnesses do not overstate their credentials, embellish their *curriculum vitae* (résumé), or outright lie about any of their qualifications. It can be expected that any competent defense attorney will scour the expert's curriculum vitae (CV), publications, and previous testimonies to come up with any inconsistencies that can be used to discredit or, better yet, disqualify the expert. It is for this reason that experts should be cautious in what they write in their CV and what they say on the record, either on the witness stand in a trial or in a sworn deposition. CVs and court transcripts from previous testimonies can be scrutinized by opposing counsel and reintroduced at trial even years after the fact.

Legal History of Expert Testimony and Forensic Evidence

How has the law dealt with determining what kind of evidence is admissible? Just like in other areas of the law, the admissibility of forensic evidence and expert testimony is rooted in *precedent*. Precedent is the principle that a theory, idea, or practice has been accepted or allowed on a previous occasion, and if everything remains the same, should be accepted again in the future. Legal precedent can therefore be found in rules, statutes, and case law.

THE FRYE TEST

The first case dealing with the admissibility of forensic evidence originated from the Court of Appeals of the District of Columbia in 1923. In the case of *Frye v. United States,* 293 F. 1013 (D.C. Cir. 1923), the Washington D.C. appeals court ruled that before a new scientific principle or discovery can be used as evidence in a court of law, it must be generally accepted within the relevant scientific community.

In the early 1920s, a man named James Frye was found guilty of murder on the basis of a new *polygraph* (or "lie detector") test, which was based on the theory that when a person lies, the blood pressure becomes elevated. The court ruled that the novel blood-pressure test had not gained acceptance, and so Frye's conviction was reversed. (In fact, polygraph evidence is still inadmissible in court today.) This rule of "general acceptance" became known as the *Frye rule* or *Frye test*. The ruling in *Frye v. United States* explained:

> Just when a scientific principle or discovery crosses the line between the experimental and demonstrable stages is difficult to define. Somewhere in this twilight zone the evidential force of the principle must be recognized, and while courts will go a long way in admitting expert testimony deduced from a well-recognized scientific principle or discovery, the thing from which the deduction is made must be sufficiently established to have gained general acceptance in the particular field in which it belongs.

This "general acceptance" standard provides the guidelines for the judge to make a determination as to the admissibility of expert testimony. General acceptance can be determined through previous testimony, peer-reviewed publications, and use of the actual procedure. Even today, almost 75 years later, the *Frye* rule continues to preside over the admissibility of forensic evidence and scientific testimony in several state jurisdictions that adopted the rule long ago.

FEDERAL RULES OF EVIDENCE

As scientific methods became more specialized, the limitations of the *Frye* rule became evident. Enacted in 1975, the Federal Rules of Evidence, under Article VII, "Opinions and Expert Testimony," describe the admissibility of scientific evidence and expert witness testimony in the U.S. federal court system. There are many jurisdictions outside the federal court system that have also adopted these rules to govern testimony by experts. The rule that applies most fittingly to forensic expert witnesses is Federal Rule 702, "Testimony by Experts," which states:

> If scientific, technical, or other specialized knowledge will assist the trier of fact to understand the evidence or to determine a fact in issue, a witness qualified as an expert by knowledge, skill, experience, training, or education, may testify thereto in the form of an opinion or otherwise, if (1) the testimony is based upon sufficient facts or data, (2) the testimony is the product of reliable principles and methods, and (3) the witness has applied the principles and methods reliably to the facts of the case.

Thus, the judge must ensure that the scientific evidence is based on a reliable foundation and that it will assist the court in deciding the issue before it. This standard is more flexible than the *Frye* rule, because it does not have the "general acceptance" requirement for admissibility.

Junk Science

With the adoption of the Federal Rules of Evidence, some legal professionals opined that the notion of scientific consensus carried less weight. Some felt that expert testimony was now allowed if it "will assist the trier of fact to understand the evidence or to determine a fact in issue," regardless of the scientific accuracy of the testimony. Some even foresaw that any iconoclasts whose views might prove "helpful" to a jury might now be allowed to testify. As a result, some courts failed to screen expert witnesses and their testimony, instead adopting a "let-it-all-in" approach. This "let-it-all-in" philosophy, in the hands of unscrupulous attorneys, however, allowed almost anybody in the witness box. Consequently, the opinions of charlatans, quacks, and imposters were sometimes afforded equal weight to those of Nobel laureates. Over the years, in fact, the legal system has certainly seen its share of junk science or pseudoscience in the courtroom.

THE DAUBERT DECISION

Prior to 1993, two main standards of admissibility existed: the *Frye* test and Rule 702 of the Federal Rules of Evidence. In 1993, the United States Supreme Court changed the decades-old law of admissibility of expert scientific testimony at the federal court level by rejecting the *Frye* test as inconsistent with the Federal Rules of Evidence, in the case of *Daubert v. Merrell Dow Pharmaceuticals, Inc.*, 509 U.S. 579 (1993).

The plaintiffs in the case were the parents of children born with birth defects. The main argument of their case was that the defendant's drug, Bendectin, was responsible for causing birth defects. The trial judge heard expert evidence from qualified scientists on both sides. The defendant's expert testified that the scientific community was generally of the opinion that Bendectin could not have caused the birth defects the plaintiffs' children had suffered. The plaintiffs' experts argued that certain data in existing chemical and animal studies could be interpreted to show that the drug might have been responsible for the birth defects. Given the rule in *Frye*, the court ruled that since only the defendant's theory was "generally accepted," the plaintiffs' expert testimony could not be admitted despite the reliability of the science behind it. Therefore, the trial judge ruled for the defendants.

The plaintiffs appealed, and the appeal reached the U.S. Supreme Court. The Supreme Court, in a unanimous decision, reversed the trial judge and sent the case back to the original court. The Supreme Court noted that Rule 702 of the Federal Rules came after the *Frye* case was decided in 1923, and decided that the "generally

accepted" rule was too restrictive. The justices outlined several considerations that the evidence had to be given by trial courts to determine its admissibility:

- Have the theories and techniques employed by the scientific expert been tested? Admissibility does not require proof that the theory or technique has been proved, however.
- Have the theories and techniques been subjected to peer review and publication? There is no requirement that the theory be accepted by the scientific community, only that it has been reviewed by people in the field. Theories that have not been reviewed or published can be excluded.
- Do the techniques employed by the expert have potential error rates? This requires an indication that the science behind the theory has been tested and the limitations of the technique can be assessed by the trier of fact.
- Are the theories subject to standards and controls governing their application? Do other scientists doing similar work employ the same standards and controls?
- Do the theories and techniques employed by the expert enjoy widespread acceptance within the relevant scientific community? While they do not need majority support in the scientific community, a substantial minority needs to support the work.

The Supreme Court ruled that the Federal Rules of Evidence, and not *Frye*, were the standard for determining admissibility of expert scientific testimony in federal cases. (This implication also reached the state court level as well, as many states adopted the federal standard.) In essence, this granted judges much more discretion in determining the admissibility of expert testimony. All of the considerations in *Daubert* do not need to be met for the evidence to be admissible; it is only necessary that a majority of the standards be met. Ultimately, this further pushed judges into the role of "gatekeeper" in determining the acceptability of new methods.

One problem with scientific expert testimony is the appearance in courtrooms of peripheral scientists who qualify as "experts." These "junk scientists" and unscrupulous attorneys collaborate to convince judges and juries to accept their frivolous claims. Judges need to require experts to represent currently accepted scientific theories rather than their own, sometimes outlandish, beliefs. After *Daubert*, it was expected that the range of scientific opinion used in court cases would expand. However, since 1993, courts have strictly applied the standards of *Daubert*, and they have generally been successful in keeping junk science out of the courtroom.

The principle in *Daubert* was later expanded in *Kumho Tire Co. Ltd. v. Carmichael*, 526 U.S. 137 (1999). The main issue in this case was that the evidence in question was from a technician and not a scientist. The technician was going to testify that the only possible cause of a tire blowout must have been a manufacturing defect. The Eleventh Circuit Court of Appeals had admitted the evidence on the assumption that *Daubert* did not apply to technical evidence, but rather only to scientific evidence. The U.S. Supreme Court reversed the Eleventh Circuit decision, saying that the standard in *Daubert* could apply to merely technical evidence as well.

While the court cases cited thus far in the chapter govern the admissibility of evidence in a court of law, it is still ultimately left to the jury to determine how much weight the evidence is afforded. Even if evidence is admitted in a case, a jury can still evaluate its strengths and weaknesses.

Types of Evidence

There are two main types of evidence used in courts of law: testimonial evidence and real evidence. *Testimonial evidence* is oral testimony given by a witness under oath. Some examples of testimonial evidence include eyewitness testimony about a person's physical description or events that the witness saw transpire. *Real evidence* consists of objects. Real evidence can consist of actual things that played a direct role in the crime, like a murder weapon or a used condom in a rape case, or items used for illustrative points, like crime-scene photos, charts, and diagrams.

In order for an item of evidence to be admitted, it needs to be authenticated. This is done when the expert witness in the witness box identifies the object. Since forensic scientists examine hundreds, if not thousands, of items of evidence in any given year, it is often difficult to remember every single item of evidence that was examined. It is for this reason that forensic scientists place their initials along with the case number directly on the physical evidence they examine so that they can recall and identify the item when questioned about it on the witness stand. One strategy of a good defense attorney might be to try to question a witness about a very similar looking object, different from the actual one they examined. It is therefore also important for forensic examiners to take detailed notes and/or sketches about the evidence including color, size, manufacturer, serial numbers, defects, and other unique identifying features. This information enables the forensic examiner to recall information about a case at trial, sometimes years after the fact.

One thing that all real evidence requires is a chain of custody. This ensures that a record is kept of every single person who handles a particular piece of evidence. Each person who has had custody of the evidence from the time it is collected through the time it is received at the laboratory, examined by the forensic scientist, and stored by the property clerk is written on the chain of custody. This ensures that when the evidence is later presented in court, the integrity of the evidence is not called into question. Just like an actual chain, the chain of custody is only as strong as its weakest link. One mistake occurring early on in the chain of custody can challenge the validity of all future testing performed on the evidence. In addition to a chain of custody, attorneys can also question the packaging of the evidence. Tamper-proof seals and proper packaging ensure that the evidence has not been tampered with nor contaminated.

Statute of Limitations

Most types of crimes have a *statute of limitations*, which is defined as a set amount of time after which the perpetrator can no longer be prosecuted. The amount of time that must pass is usually based on the severity of the crime, but there are exceptions, and even some crimes that have no statue of limitations. The most common example of a crime with no statue of limitations is murder, which society has deemed such a heinous crime that it is worth prosecuting the suspect even years after the fact.

The rationale behind the statue of limitiations is that after a long period of time has elapsed following the occurrence of a crime, memories of victims, witnesses, and suspects start to fade. In many jurisdictions, for example, the statute of limitations for rape ranges between 5 and 10 years. This means that even if the identity of the suspect is known, if charges are not brought within the set period of time, the suspect cannot be criminally prosecuted. Charges can be brought against the identified suspect even if the suspect cannot be found and arrested, as long as the charges are brought before the statute of limitations expires. This prevents the suspect from claiming a statute of limitations defense in the future.

The problem exists, however, of what to do if the identity of the perpetrator is unknown. How can a person be *indicted*, or charged with a crime, if their identity is unknown to prosecutors? Recently, courts have allowed prosecutors to indict unknown suspects, or "John Does," based on their DNA profiles. DNA evidence (discussed in Chapter 7) is oftentimes left behind in a sexual assault. Even though the DNA contributor may be unknown to the victim, such as in a rape case involving a stranger, a DNA profile is a unique description of an individual, so it can be used to identify a person to the exclusion of everyone else. Essentially, then, the DNA profile can be indicted. If the suspect is later apprehended after the expiration of the statute of limitations, he or she can still face prosecution since they have already been indicted.

Direct Examination

Under direct examination by an attorney, an expert witness conveys to the trier of fact how he or she reached an opinion based on scientific facts. This is an opportunity for a good expert witness to develop rapport with the judge or jury. They will speak slowly and clearly to the trier of fact, facing them when answering questions instead of the attorney asking the questions. After all, it is not the attorney who decides the verdict. Although answers should be brief on direct examination, the expert witness is allowed more room to converse with the attorney than while under cross-examination.

With the ever-increasing technology in forensic laboratories, it cannot be assumed that attorneys will always be able to grasp the scientific principles that are utilized by forensic experts. Many attorneys do not have a strong background in science. Therefore, an attorney typically requests a meeting with the forensic expert prior to the expert's testimony in a deposition or trial so that the expert can go over their case notes and conclusions with the attorney. This "pretrial" meeting enables the expert to teach the attorney about the expert's area of expertise so that the attorney knows what questions to ask the expert and how to formulate the questions to solicit the information that needs to be conveyed to the trier of fact.

A good expert witness will also be able to anticipate possible questions that could be raised by a defense attorney under cross-examination, and address these questions during direct examination. This can potentially prevent what appears like the expert being caught off guard and can win the expert favor in the eyes of the jury. For example, if an expert witness knows of an error or a problem with the work on a case, it is usually better to bring this up during direct examination, instead of waiting for opposing counsel to make the inquiry.

Cross-Examination

The justice system in democratic countries allows the accused to confront witnesses called to testify against them. This is a basic right afforded to all and allows the accused the chance to offer a defense and attempt to raise possible doubt in the minds of the judge or jury regarding the culpability of the accused.

With the multitude of courtroom television programs, cross-examination is often seen by the public as the ultimate drama between witness and attorney. In reality, however, an unbiased expert witness should not change tactics under cross-examination and should continue to answer questions with the same demeanor as under direct examination. A good expert witness remains calm and is in control of their testimony at all times, never becoming hostile or confrontational with opposing counsel. They listen carefully to the questions being asked of them and ask counsel to define or clarify certain terms if their meaning is unclear. The expert witness should keep in mind when testifying that they are in the courtroom to educate the trier of fact and not to be an advocate for either side.

It is also important for forensic experts to maintain their impartiality, in spite of being called as an expert by either the plaintiff or the defendant. Even though the justice system is an adversarial one, if an expert called by the prosecution has forensic evidence that can help the defendant, it is incumbent upon the expert to inform the court of this information. It is for this reason that any fees charged by the expert for testifying should be paid in advance, and should not be contingent on the substance of their testimony or the outcome of the trial.

Discovery

Discovery is a legal term that refers to the right of one side of a case to question the other side. In criminal cases, the prosecutor is required to turn over to the defense any witness statements and any evidence that might tend to exonerate the defendant. Depending on the rules of the court, the defendant may also be obligated to share evidence with the prosecutor. Forensic scientists should have some familiarity with the discovery rules in their jurisdiction so that they can advise an attorney about anything that might need to be turned over to the opposing side.

Sections F and G of Rule 16(a)(1) of the United States Federal Rules of Criminal Procedure, which many jurisdictions have adopted either in whole or in part, state:

(F) *Reports of Examinations and Tests.* Upon a defendant's request, the government must permit a defendant to inspect and to copy or photograph the results or reports of any physical or mental examination and of any scientific test or experiment if:

    (i)   the item is within the government's possession, custody, or control;

    (ii) the attorney for the government knows—or through due diligence could know—that the item exists; and

    (iii) the item is material to preparing the defense or the government intends to use the item in its case-in-chief at trial.

(G) *Expert Witnesses.* At the defendant's request, the government must give to the defendant a written summary of any testimony that the government intends to use under Rules 702, 703, or 705 of the Federal Rules of Evidence during its case-in-chief at trial. If the government requests discovery under subdivision (b)(1)(C)(ii) and the defendant complies, the government must, at the defendant's request, give to the defendant a written summary of testimony that the government intends to use under Rules 702, 703, or 705 of the Federal Rules of Evidence as evidence at trial on the issue of the defendant's mental condition. The summary provided under this subparagraph must describe the witness's opinions, the bases and reasons for those opinions, and the witness's qualifications.

It is, therefore, important that the forensic expert witness keep their attorney informed about all aspects of the testing performed in their case so that it can be disclosed to the other side if requested. In addition to this Federal Rule of Criminal Procedure, discovery is also governed by two Supreme Court cases. *Jencks v. United States*, 353 U.S. 657 (1957), which became codified as the Jencks Act, regulates discovery and disclosure statements of prosecution witnesses. In *Brady v. Maryland*, 373 U.S. 83 (1963), the Supreme Court declared that, regardless of the good faith or bad faith of the prosecution, the suppression of evidence favorable to the defendant violates due process where either guilt or punishment is dependant on the evidence.

Limitations of Forensic Science

It should be noted that while forensic evidence and expert testimony can provide very useful information to the trier of fact, it is not the only factor that should be considered in deciding a verdict. As is true for most areas of science, forensics has its limitations. Although forensic evidence can be very helpful to the judge or jury in reaching a decision, typically it is only one of many factors the trier of fact must consider.

Quiz

Refer to the text in this chapter if necessary. Answers are located in the back of the book.

1. The role of judge as "gatekeeper" is best summarized by:
   (a) *Frye*
   (b) Daubert
   (c) Rule 702 of the Federal Rules of Evidence
   (d) The criminal justice system

2. As an expert witness in the courtroom, you are allowed to:
   (a) Only state matters of fact pertaining to the case at hand
   (b) Only state opinion pertaining to the case at hand
   (c) Give broad generalizations about the possibilities of how a crime occurred
   (d) State your opinion in light of the scientific testing done in the case

3. As an eyewitness in the courtroom, you are allowed to:
   (a) Only state matters of fact pertaining to the case at hand
   (b) Only state opinions pertaining to the case at hand
   (c) Give broad generalization about the possibilities of how a crime occurred
   (d) State your opinion in light of the scientific testing done in the case

4. The best evidence from a crime scene to present in court from the following is:
   (a) A similar gun to that which was used in the crime
   (b) A physical description of a .357 Magnum
   (c) The actual murder weapon, a .357 Magnum
   (d) A photograph of the murder weapon

5. The trier of fact can be:
   (a) The expert witness
   (b) The prosecutor
   (c) The judge
   (d) The defense attorney

6. The hearsay rule:
   (a) Permits experts to render an opinion
   (b) Does not apply to expert witnesses
   (c) Allows secondhand testimony at trial
   (d) None of the above

7. The process in which an expert witness is qualified on the stand is:

   (a) Amicus curiae

   (b) Habeas corpus

   (c) Duces tecum

   (d) Voir dire

8. A defense expert witness:

   (a) Can expect to receive a bonus if their side wins

   (b) Can charge reasonable fees for their testimony

   (c) Cannot charge for their testimony

   (d) Is exempt from reporting any income earned from testifying on their tax return

9. Which case deals with discovery issues?

   (a) *Brady v. Maryland*

   (b) *Marbury v. Madison*

   (c) *Plessy v. Ferguson*

   (d) *Kramer v. Kramer*

10. If evidence falls under discovery rules and is not turned over:

   (a) The case can be dismissed

   (b) A mistrial can be declared

   (c) Witness testimony can be stricken from the record

   (d) All of the above

CHAPTER 3

Crime Scene Investigation

The scene of the crime represents the starting point of any forensic investigation. The crime scene is a representation—a snapshot in time—that serves as a record of the actions and events of the crime. Any forensic science information that is related to the crime is there to be discovered. Careful examination of the scene may yield physical evidence, which can be examined in a crime laboratory to reconstruct how the crime occurred. Because so much information may be gleaned from the crime scene, it should remain undisturbed and in its original condition until trained crime scene investigators and forensic specialists can be brought into the investigation. Of course, how complete and thorough the investigation is in any particular criminal case depends on the nature of the crime; burglaries may not get the same attention as murder cases, for example. This is due to resource limitations. Police should always consider, however, that seemingly innocent cases and even apparent accidents might actually be more serious crimes masquerading as innocuous events.

Many "accidental deaths" turn out to be murders that were covered up by a skillful perpetrator.

A uniformed police officer is generally among the first responders to arrive at the scene of a crime. Paramedics, fire fighters, and even family members or friends sometimes arrive before the police, and may try to render first aid or resuscitation in violent crimes. Police departments work closely with paramedics and firefighters to teach them the basic skills of crime scene preservation so that physical evidence is not unduly compromised or destroyed. The close cooperation of all personnel who may become involved in a criminal investigation is important to the successful outcome of cases. Errors and omissions, often accidental, can turn straightforward cases into difficult ones. Sloppy work at the crime scene is not in anyone's best interest, except perhaps for the perpetrator.

No two crime scenes are exactly the same. Human nature is such that there is an infinite number of variations on how criminals may commit their crimes. Yet, experienced crime scene investigators often develop a degree of intuition on what is important at a scene. There is no checklist to follow when examining crime scenes or investigating crimes. There are, however, certain protocols to follow to collect and document evidence properly so that it is admissible in court and workable in the crime lab. These guidelines are well understood and practiced by crime scene investigators and police worldwide and are the basis of sound forensic practice.

The actions of the first police officer to respond to the crime are critical to the rest of the investigation. The police officer's sole duty is to protect the crime scene from change. Except in minor criminal investigations, the first officer is usually not expected to collect and document evidence. Among the officer's duties are to cordon off the scene to prevent unauthorized persons from passing through the scene, so that evidence is not moved or contaminated. The responding officer also must ensure that the witnesses and suspects are separated so that they do not have a chance to compare observations or conspire to change their versions of what happened during the commission of the crime (see Figure 3-1).

Depending on the nature of the scene, multiple perimeters might need to be established by the first responding officer to minimize access to various sections of the crime scene. The outermost perimeter can be used to keep the general public and curious onlookers at bay, while the next level of perimeter might be set up for press access only. Past that, there might be other perimeters to allow only the most critical investigators access to the actual crime scene. To ensure the security and integrity of the crime scene, a log is used to maintain a list of those persons and the time that they entered and exited the scene. Almost everything that follows in the investigative process may be tracked back to effectively conducting these tasks.

Figure 3-1 Cordoned-off crime scene

General Crime Scene Procedure Guidelines

Protecting the crime scene can be accomplished in a variety of ways. The most common is the use of commercially available yellow crime scene tape to cordon off the area. This is especially useful outdoors. The tape can be easily wrapped around trees, telephone poles, or other stationary objects and defines the area. Barricades, police cars, or any other convenient items at hand may be used to designate the crime scene and keep nonessential people out of the area. The reason for keeping people out is simple: too many people milling about an area will introduce changes to the scene and destroy evidence. The simple rule is: the fewer people in the crime scene proper, the better. This also applies to high-ranking police officials who have no business at a crime scene other than to see what is going on.

Locard's Exchange Principle, discussed in Chapter 1, is the primary reason for keeping the crime scene pristine and keeping nonessential people outside of the crime scene. Simply put, Locard's Exchange Principle states that every contact leaves a trace. It is not possible to enter an area without adding something or taking something away (generally of a microscopic nature). Thus, when too many people are at a scene, changes occur, and those changes may result in loss or damage to key trace evidence. However, even if only appropriate people are present inside the crime scene, Locard's Exchange Principle still applies to them, and every effort must be made by police, investigators, and forensic personnel to minimize their interaction with the scene. For that reason, it is important for crime scene investigators to wear gloves, Tyvek suits, and foot covers to minimize the amount of trace evidence that they inadvertently bring into and take out of the crime scene.

One thing that police officers should consider is that victims and suspects sometimes clean up the scene before the police arrive. This may be done to cover up the crime, or simply to clean up a mess left behind at a crime scene. When crime victims call the police, the dispatcher should warn them not to clean up or move anything. After the scene is secured by the first officer to arrive at the scene, and witnesses and suspects are removed, detectives will interview those individuals to determine if anything was moved or handled at the scene.

While waiting for detectives to arrive, the first officer on the scene should do the following:

- Conduct a cursory search of the scene for other victims and suspects.
- Render first aid as needed—the first priority is to save lives!
- Write down names of witnesses and other persons present at the scene.
- Note who was at the scene when he or she arrived.
- Establish the facts of the crime.
- Keep any suspects and witnesses separated and instruct witnesses not to discuss the events.
- Make sure not to discuss the crime with witnesses or bystanders.
- Listen attentively.
- Protect evidence that could easily be destroyed.
- Be prepared to brief detectives.

For similar reasons concerning scene contamination, suspects should never be taken through the crime scene. This imperative may be overlooked if the scene has been examined. A suspect, however, may change some aspect of the scene. The detective responsible for the investigation should have the final say in such situations.

A question often asked in the context of crime scene investigation is this: how much evidence should be collected? Is there such a thing as collecting too much evidence? Most forensic labs in the United States have backlogs, because their resources are not sufficient and they are unable to do everything. Because of this, labs often triage cases, working on the most serious ones first before moving on to less serious ones. Crimes against persons, such as murders, rapes, and assaults, generally take precedence over property crimes such as burglaries.

Experience is clearly a determining factor in answering the earlier question. Crime scene investigators and detectives develop a knack for determining how much evidence to collect. Having said this, it is important to constantly ask, "Have I overlooked something? Is it possible that there is some other circumstance that I am missing as I 'read' this crime scene?" In addition to deciding what to collect, investigators also should have some familiarity with the capabilities of the crime lab they use. Throughout any investigation, communication between the detective, crime scene investigator, forensic scientist, coroner/medical examiner (in murder cases), and prosecutor is important.

Recognizing physical evidence helps the crime scene investigator develop a plan of how to, and in what order, to collect evidence. Generally, more fragile evidence is gathered before more robust evidence. The crime scene will be photographed to note the location of evidence and its relationship to the area. Fingerprints are usually collected next. Latent prints, those that are not readily visible, are easily destroyed and even missed.

In addition to developing a plan of action, recognizing evidence at the scene and its relationship to the location and the crime helps investigators reconstruct the manner in which the crime occurred. Crime scene reconstruction is often helpful in the investigation and subsequent prosecution of the crime because it may be used to impeach a defendant's testimony and explain how the crime occurred.

Careful observation is a key to this process. Crime scene investigators and detectives should be careful to recognize both the ordinary and the unusual at a crime scene. The following is a list of some of the observations that may be important in a case:

- Were the windows open or shut? Were they locked?
- Were the doors open, closed, or locked? Was the key in the lock?
- Were the lights on or off? If on, which lights were on?
- Were the shades, shutters, or blinds open or closed?
- Were there any odors such as cigarette smoke, gas, gun powder, or cologne?
- Were there any signs of activity such as meal preparation, or dishes in the sink?

- Was the house well kept or in disarray?
- Were there any date or time indicators such as mail, newspapers, "sell by dates" on food items, spoiled food, unwound clocks, items that should have been hot or cold but were at room temperature?
- What were the exterior lighting conditions and weather at the time of the crime?
- Was there anything that seemed out of order or just not right?

Documentation of the crime scene search and evidence collection is critical to any criminal investigation. Documentation includes notes, photographs, and sketches of the crime scene. In major cases, video recordings may also be made. There are two main purposes for detailed, contemporaneous, and proper documentation:

- Documentation helps investigators recall all the work done at this phase of the investigation. Cases often take months, and sometimes years, to solve.
- Documentation helps to establish the *chain of evidence*. In order for physical evidence to be admitted to a court, investigators must be able to document everything that happened to that evidence. Notes, photographs, and sketches help to establish the chain of custody (see Figure 3-2).

Detailed, permanent notes in ink made in a bound book are a must. Notes should be made as the events unfold, in the order that they occur. They should be detailed and include all actions. Notes should be complete and thorough; they should be clearly and legibly written. (Sloppy notes or those that do not clearly state what the investigator meant will be subject to misinterpretation and can even point to a shoddy investigation.)

Even negative or unexpected conditions, such as the absence of bloodstains or a lit light in the daytime, should be included. Notes should be as specific as possible. When the location of evidence is noted, a description such as "on the kitchen floor, 10 inches west of the east wall and 2 feet south of the north wall" should be used. Unclear statements should be avoided.

Notes   Photographs   Sketches

Figure 3-2   The principle ways to document a crime scene are thorough, clearly written notes, adequate photographs, and crime scene sketches.

Case notes, sketches, tape recordings, and photographic negatives should be stored in a place where they will not deteriorate. They should be placed in a case folder and retained for as long as the department's policy indicates. It is not uncommon for cases to be reopened years later and sometimes by a new investigator.

Crime scene notes should include the following information:

- Date and time the police were called
- The address of the crime scene and a description of the area
- The name of the person who called the crime scene investigator to the scene
- Names of all officers, witnesses, investigators, and other personnel at the scene
- Names of the persons who conducted the crime scene search, including persons who took photographs, fingerprints, and sketches and collected evidence
- Weather and lighting conditions
- Description of the location of the body
- Location of any evidence found during the investigation
- Description of the interior and exterior of the crime scene
- Description of the outside area of the scene
- Date and time the crime scene investigation was concluded

The adage that *a picture is worth a 1000 words* is certainly true when describing the documentation of the scene of a crime. Photographs, like notes, are designed to refresh the investigator's memory and explain to juries what the scene looked like. Photographs should be taken close up, at medium distance, and far enough to get the overall scene. Pictures help to show the appearance of the scene at the time the police arrived and the relationship of various objects at the scene to one another. Photographs should be taken with two rulers (scales) in the frame and at right angles to the items being photographed. A "chain of custody" will be required to permit photographs to be admitted into evidence in court. The name of the photographer, the date and time the photos where taken, and other pertinent details must be included in a photo log to establish this chain.

Sketches are another way to describe a crime scene. Sometimes photographs may contain too many details and make it hard for a jury to understand the relationship of various elements of the crime scene. A simple sketch can help clarify important elements of the scene. Careful measurements are an important part of any crime scene sketch. As with photographs, a chain of custody is needed. It is not unusual for the sketch maker and photographer to be called to court to testify about the making of the photos and sketches.

Crime scene searches are thorough, systematic procedures to look for physical evidence. Aimlessly wandering about an indoor or outdoor crime scene is not at all fruitful and generally a wasted effort. Generally, crime scene investigators choose a systematic approach that works for them. There are several types of search patterns: lane search, grid search (see Figure 3-3), spiral search, etc.

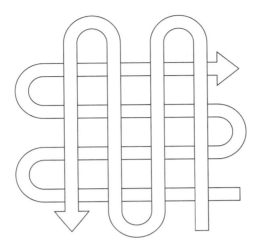

Figure 3-3   Search patterns

The purpose in using a search pattern is to conduct a thorough search of the area to collect and document all the possible evidence available. The type of search to use depends on the circumstances and nature of the scene and what works best in a given situation. Whatever the procedure used, the goal is to conduct a complete search of the location. One consideration that is sometimes forgotten is that crime scenes are three-dimensional spaces. Investigators must look for evidence at the ground level and above ground level. Sometimes important clues may be present on walls, ceilings, or even roofs.

Some experienced investigators first spend time to take in the overall scene before going on to search for individual items of evidence. This process can give the investigator or detective a sense of the dynamics of the scene and help to determine how the crime was committed. Experience often plays a role in searching the crime scene. Conducting crime scene searches with senior personnel can help to train less experienced investigators.

Crime scene investigators should consider that criminals sometimes do things without thinking. Sometimes fingerprints are left in places that are so obvious that they might be missed. For example, the criminal may have used the toilette, had a drink from a glass, or opened the refrigerator for a snack. In criminal investigations, literally anything is possible and investigators should not assume that because a scene is neat and clean that evidence will not be found. Similarly, a scene that is in disarray may also contain important evidence to a case.

Crime scene investigators sometimes begin the examination of a scene by trying to imagine what occurred and how the perpetrator may have acted during the commission of the crime. This can be helpful as long as the investigator does not permit herself to become too invested in one theory. Sometimes, information may come to light to cause a total change of an initial hypothesis. Tunnel vision, a phenomenon where the investigator becomes invested in only one scenario and ignores other possible scenarios, must be avoided.

During the entire search process, the continued protection and security of the crime scene cannot be ignored. This is particularly a challenge during cold and inclement weather, when police would prefer to congregate inside an indoor crime scene to keep warm and dry. Of course, the downside of that situation is the chance of disturbing or destroying important trace evidence.

While the crime scene search is in progress, investigators should take contemporaneous notes detailing all of their activities. When an item of evidence is observed, it should be photographed in place, with and without a ruler or scale. Careful measurements should be taken to be included in both the notes and crime scene sketch in order to show the location of the evidence in relation to the rest of the crime scene.

As noted earlier, there are several typical schemes that may be used to search a crime scene. The grid pattern is popular. Sometimes a spiral pattern may be appropriate. In this case, the search would begin at a central point of the scene. For example, the location of a body in a murder investigation might be such a point. From that point, crime scene investigators would travel in a spiral, making each pass of the remains wider than the previous one. No matter what technique is used, a careful and thorough search is the key for any successful crime scene investigation.

Outdoor scenes require a different approach from indoor scenes. Of obvious concern to the police is the open access of the scene to the news media. There is a critical balance between the public's access to information and the confidential nature of a criminal investigation. The police cannot successfully investigate a crime if all information is public. With this in mind, it is common for police to attempt to block off the scene from outside observation, including overhead observation from helicopters.

Crime scenes on public streets or highways present other challenges. Traffic flow becomes an added consideration. It is simply impractical to close down a roadway to traffic for a long period. Thus, crime scene investigators in such cases need to move quickly to complete their tasks and try not to give up quality for the sake of speed.

For large outdoor scenes in an open field, searching for evidence becomes the exercise of locating the "needle in the haystack." To accomplish such searches, lane or strip searches are sometimes used, where investigators slowly walk in a straight line following a lane. Another successful approach is a grid. Plots, perhaps 6 feet square, are staked out and a person is assigned to carefully examine one small plot at a time. These searches can be laborious, but can lead to critical evidence.

Another concern for investigators to worry about in outdoor scenes is the weather. Rain, snow, and rising or falling temperatures may have some effect on how a case is handed. For example, if shoe or tire impression casts need to be made, weather changes can have a deleterious effect and appropriate actions are required, such as erecting a canopy over the scene.

Reference points at outdoor scenes present a further difficulty. In a large outdoor area, it is necessary to find permanent reference points—for example, a large rock, a recognizable tree, or some other landmark. A surveyor may be required to assist in locating evidence in a drawing or diagram. The Global Positioning System (GPS) may provide some assistance, but current technology may be able to pinpoint a location to within only a few feet. In addition to GPS devices, an excellent piece of equipment to have at an outdoor crime scene is a *total station*. A total station integrates the functions of a *theodolite* (an instrument used for the measurement of horizontal and vertical angles in surveying), an electromagnetic distance meter (EDM) for measuring distances, and digital information recording. A total station can be used to *triangulate*, or determine the distances between certain objects, in a large outdoor crime scene.

Whether the scene is outdoors or indoors, it is important to try to determine the path the perpetrator used to enter and leave the location. In crimes that begin as a break in, important evidence is often present at the entry and exit of the scene. These may include fingerprints, tool marks, blood, footprints, and various kinds of trace evidence. The specific location of a body in a murder investigation is another location that requires particular care. If a struggle took place, transfer evidence between the victim and assailant and actual scene may yield important clues.

Crime Scene Reconstruction

Recognition, identification, documentation, and collection of physical evidence at crime scenes are elements of the investigative process. Forensic investigators are sometimes called on to perform another process, known as *crime scene reconstruction*. Crime scene reconstruction takes all the observations gleaned at the crime scene and the conclusions of the autopsy and crime lab reports and attempts to reconstruct the circumstances of the events leading up to and culminating with the criminal act. The process, reminiscent of how the fictional crime detective Sherlock Holmes might approach an investigation, is in reality the use of deductive logic and inductive reasoning. By synthesizing all the information available, the investigator attempts to "put it all together" into a logical, coherent fashion.

One of the many techniques used in crime scene investigation is bloodstain pattern recognition, sometimes also know as blood spatter pattern interpretation. The examiner, familiar with the way liquid blood reacts under different circumstances, can determine with a degree of reliability how bloodstains left at the crime scene occurred. For example, gunshots to the head cause high-velocity spatter patterns

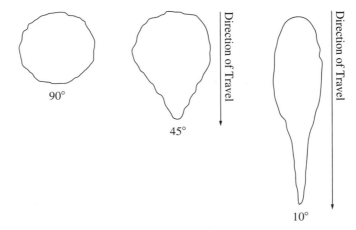

Figure 3-4 Blood droplets patterns

that are characterized by a fine mist left on a wall near the target. Medium-velocity spatter, such as that caused by a baseball bat, leaves a different pattern with somewhat larger droplets.

Blood dripping on surfaces can also tell a story. Depending on the height above a surface and the nature of the surface, blood may look like round drops or have scalloped edges around the circumference of the drop. Blood striking a surface at an angle will leave teardrop-shaped stains. All these patterns are reproducible and characterized through experiments (see Figure 3-4). Figure 3-4 depicts how blood appears when dropped onto a surface at various angles: 90°, 45°, and 10°. The tail on the droplet indicates the direction of motion.

Collection of Physical Evidence

Two considerations are critical for evidence collection and preservation for subsequent submission to a forensic science lab and ultimately to the court: legal considerations and scientific considerations. For evidence to be legally admissible, the *chain of custody* of the evidence submitted to the court must be proved. The chain of custody provides documentary proof that the evidence submitted to the court is the same as that collected during the criminal investigation and that any changes or modifications to the evidence can be explained and documented. Generally, evidence is packaged and tagged in such a way as to demonstrate that what arrived in court is the same as that collected.

For some cases, such as collecting DNA reference samples from suspects or certain physical evidence located out of sight in a person's residence, a court order or search warrant is legally required. If an investigator is uncertain about a specific instance, it is always good practice to contact prosecutors for their advice. It is generally better to take a conservative approach and not compromise the admissibility of evidence at the trial. Judges can be harsh if there is a clear violation of search and seizure laws when collecting evidence. Critical items of evidence in a case can be thrown out of court if the constitutional rights of the accused are not honored.

Scientific considerations are important in lab testing and legal arguments concerning the integrity of the evidence. Contamination and preservation of evidence are the critical factors. For example, packaging items of evidence and reference samples in the same containers, leading to contaminated evidence, would result in a meaningless lab conclusion. Similarly, packaging wet, bloody items in plastic, which creates an environment for mold growth, renders the evidence useless.

Collecting evidence ultimately depends on the type and condition of the evidence. Any "how to" list would become rather long, so the following condensed list is designed to give you an understanding of some of the factors to consider:

- Items with fingerprints or dust prints should be packed so that they do not come in contact with each other or the package material.
- Bullets and evidence containing tool marks should be protected from damage.
- Clothing containing blood, semen, saliva, etc. must be packaged separately so that the stains are not disturbed. Damp items must be air dried and placed in paper wrappings, instead of plastic.
- Victim's and suspect's clothing should be packaged separately.
- Hairs on clothing should be removed, labeled, and packaged separately so that they are not lost.
- Guns and rifles should be tied down in cardboard boxes without further wrapping. Postal regulations should be reviewed before shipping weapons.
- Bullets and cartridge casings should be packed separately in nonabrasive packaging material.
- Shipping requirements should be reviewed before sending live ammunition and explosives.
- Tissue, whole blood samples, and other toxicology specimens require special handling and packaging. Shippers and the lab conducting the tests should be consulted for appropriate packaging.
- Drugs and other controlled substances should be packed in cushioned containers to avoid breakage or leakage.
- Charred paper, which is especially fragile, is best hand carried to the crime lab, if possible.

Trace evidence is, as its name implies, small and often microscopic. Trace evidence may consist of hairs, fibers, soil, debris, paint, broken glass, and so forth. Searching for and collecting trace evidence is sometimes a daunting task at a scene.

In the past, specially adapted vacuum cleaners where used to collect trace evidence. The problem with that technique is that too much material is collected. A technique often used today is tape lifting, similar to the procedure one might use to remove lint from clothing. Cellophane tape, sticky side out, is patted over a surface to remove only the surface layer of debris. The tape may be applied to clear plastic or glass sheets and packaged for shipping to a lab.

Control samples are often required for comparative analysis of trace evidence. For example, if fibers are found on a suspect's clothing, it is essential to obtain fiber exemplars or control samples from the scene or suspect's clothing to compare the evidence samples. Control samples must be package carefully with the same considerations for chain of custody requirements and contamination issues that are given to any other sort of evidence.

An important element for the proper collection of physical evidence is documentation, both in field notes and on the actual packaging materials. Notes including the name of the collecting official, along with initials or signatures, dates and times items were collected, locations of the evidence at the scene, and seals are all important to prove that a chain of custody was maintained.

If the U.S. Postal Service or commercial carriers are used to ship evidence to crime labs, the same concerns about contamination, deterioration, and the chain of custody of evidence are still in play. Evidence should be shipped by certified mail with proof of delivery. If items require refrigeration, appropriate shipping procedures must be used. Fragile items must be appropriately packaged and marked to ensure that they are received in good order.

All of the early aspects of an investigation, including the crime scene search, documentation of the scene, collection of evidence, packaging specimens, and transporting specimens to the forensic science lab, are important to the outcome of an investigation and subsequent trial. Thoroughness pays off, as many an experienced detective and crime scene investigator can verify.

Quiz

Refer to the text in this chapter if necessary. Answers are located in the back of the book

1. The best way to transfer trace evidence from the crime scene to the lab is to use a:

    (a) Coin envelope

    (b) Regular postal envelope

    (c) Tape lift

    (d) Cardboard box

2. The "tail" of a blood spatter droplet, if present, points:

   (a) Back to the origin of the spatter

   (b) Back to the weapon that created the spatter

   (c) Away from the origin of the spatter

   (d) Away from the weapon that created the spatter

3. When photographing evidence details, what is the minimum number of scales that should be used?

   (a) One

   (b) Three

   (c) Scales should not be used in photography.

   (d) Photography should not be used for detail documentation.

4. Wet blood stains are best preserved by:

   (a) Freezing

   (b) Drying

   (c) Irradiating

   (d) All of the above

5. To reduce contamination at a crime scene, which of the following personal protective devices can be worn?

   (a) Gloves

   (b) Foot covers

   (c) Tyvek suits

   (d) A combination of A, B, and C

6. Documentation of physical evidence should be:

   (a) Legible

   (b) Contiguous

   (c) Detailed

   (d) All of the above

7. In packaging of evidence at a crime scene, wet evidence should be:

   (a) Dried at the scene and packaged in plastic

   (b) Dried at the scene and packaged in paper

   (c) Dried at the lab and packaged in plastic

   (d) Dried at the lab and packaged in paper

8. What type of hazardous conditions might be encountered at crime scenes?

   (a) Biological

   (b) Chemical

   (c) Environmental

   (d) All of the above

9. Which of the following is not documented at a crime scene?

   (a) Names of the personnel at the scene

   (b) The film speed at which the photos are taken

   (c) The ambient temperature of the scene

   (d) All of the above are documented at a crime scene.

10. The ideal trace collection method for a large area is:

   (a) Scraping

   (b) Tape lifting

   (c) Fingers

   (d) Both A and C

CHAPTER 4

Forensic Photography and Documentation

The overriding purpose of any forensic investigation is to apply the scientific method to the investigation of a crime, and in the process collect the facts, results, and proof to serve as evidence that can be presented in a court of law. One of the most important and influential types of evidence that can be presented in court is photographic evidence and other types of visual forensic documentation. Anybody, whether or not they are a forensic scientist, can verbally describe a scene or the conditions of a crime, but showing a photograph or a video of that scene to the court is far more valuable, because, as the saying goes, "a picture is worth a thousand words." By using correct forensic documentation at a crime scene or in a laboratory, the forensic scientist is also able to remove any inherent bias (either intentional or unintentional) in their presentation to the court. The proper use of a photograph should go hand in hand with the forensic case notes, crime scene sketches, and other written forms of documentation in the case, and all of the documentation should serve to highlight, explain, and clarify all of the evidence in the case. In a courtroom setting, though, the photograph, video, or other graphic documentation provides that necessary visual aid, and brings the court into that critical moment in time that was captured and preserved in the proper forensic manner.

So what differentiates normal photography from forensic photography? Well, to begin with, the basics are all the same. A camera is used to collect and focus light onto a film surface, which, depending on the settings of the camera and the type of film used, captures some aspect or detail of what is being photographed. In forensic photography, though, the goal of each photograph is to reveal, highlight, and document the evidence for the case in as much detail as is necessary. Because no two crimes or crime scenes are ever the same, there is also no single manner or way in which photography "must" be used at such scenes. A forensic photographer must be ready to document several different aspects of a crime in many different ways. Also, depending on the size of the forensic crime scene unit or forensic lab, there might be a single person or a team of people dedicated to photography, but all good forensic scientists should know the basics of forensic photography and be able to apply that knowledge in all types of circumstances and locations.

Documentation of Physical Injuries

One location that forensic photography is useful is a hospital or autopsy room, where documentation of wounds, bite marks, lacerations, and abrasions must be freshly documented before they are treated (on a living victim) or dissected further (as is done in an autopsy). Because these photographs are used to document pattern injuries, a high-quality camera should be used so that fine detail of the wounds is not missed.

Unfortunately, many emergency rooms do not have the time or budget to retain a full-time forensic photographer or train nurses and doctors in the proper aspects of photography. Even worse, such emergency rooms routinely use Polaroid instant cameras because of the ease of use and rapid development of the photograph. These cameras usually have poor film quality for capturing high-detail photographs, and they are also incapable of close-up photography that is necessary when dealing with bite marks, which are commonly encountered as evidence in crimes.

The American Board of Forensic Odontology Bitemark Scale

In the past, forensic documentation of bitemark evidence was troublesome, not because the photos were of poor quality, but rather because teeth, three-dimensional objects, were being compared with two-dimensional photos. If the photos were not taken at directly 90 degrees to the bitemark evidence, there would be a shift in the measurement of the bitemark in one direction, and without a reference in the photo to measure this shift, a comparison of the bitemark to a suspect's teeth could not be made. This changed with the creation of the American Board of Forensic Odontology (ABFO) No. 2 scale. The ABFO No. 2 scale is an L-shaped ruler with crosshair circles at the ends of the "L" and in the corner, as well as a full standard and metric ruler (see Figure 4-1).

When placed next to a bitemark and photographed, the scale not only provides the necessary measurements in both directions, but the crosshair figures can be measured in the resulting photograph and the angle at which the photograph was taken can be determined. If, for example, the photograph was taken at an exact 90-degree angle, the crosshair circles will be perfect circles in the photograph. As the camera is tilted, perhaps to a 45-degree angle, the crosshairs in the photo will be distorted, appearing as ovals. The geometric change in the circle directly corresponds to the angle of the camera, and the bitemark evidence can therefore be calculated at its true measurements and compared against the suspect's teeth in the case.

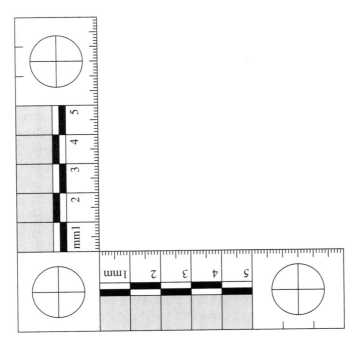

Figure 4-1 The ABFO No. 2 scale

Documentation of Forensic Evidence

A second common location for using forensic photography is the forensic lab. As evidence is opened and examined in the lab, it is necessary to document the state and condition of the evidence, as well as the packaging that it arrived in. As the evidence is examined, close-up photographs may be necessary to further document minute details of the evidence or highlight areas that are selected for testing. Also, as testing proceeds and swabs, cuttings, and scrapings are taken from the evidence, it will be necessary to document the physical cuts, scrapes, and areas on the evidence that were tested or altered during the examination. These photographs will then be incorporated into the lab notes for the evidence, and might even be labeled by the casework analyst, perhaps to point out particular stains, marks, indentations, or minutiae that identify and individualize that evidence. Forensic photography is of particularly unique importance in a fingerprint analysis lab, where the photograph of the fingerprint itself will be scanned into a computer for comparison to the AFIS national fingerprint database.

Forensic Photography at the Crime Scene

While there are other locations apart from hospitals, autopsy rooms, and labs that require the use of a forensic photographer, the most prominent location for forensic photography is the crime scene. Because time is always a factor at a crime scene, it is important for the scene to be photographed in its entirety early on and then sequentially as the scene is processed over time. That way, there will be a standing record of the progress of the work that was done at the scene as well as photographic documentation of the changes affected on the scene by the presence of the crime scene personnel.

Photography at the crime scene is also necessary for the documentation of the evidence in its original state before it is collected and removed from the scene. Even though the evidence will be further documented later on, the relationship of that piece of evidence must be established at the scene with a series of photographs. Those photographs are known as *establishing* photographs, and serve to cover the scene from a wide angle or distance and place the scene in relationship to a landmark, address, or some other general location. For example, for a crime scene at a house, a series of establishing shots would be taken of the outside of the house and would include the house address, the street signs and addresses, as well as the conditions of the exterior of the house, the street, the lawn, and so forth. Moving inside the house, more establishing shots can be taken of entire rooms or floors of the house to capture as much of the scene as possible in one overall photograph. Although lighting, the use of flashes, and the basics of photography are necessary for these photos, no scales or references are needed in these establishing shots. Additionally, there should be a brief written description of what is being photographed for each photo recorded in a photo log.

The second type of forensic photographs taken at a scene are *midrange* shots, which in effect "close in" on the actual details of the scene without losing the ability to place the evidence in relationship to the location. For example, if the establishing shot of a room was taken from a corner, the midrange shot might be taken closer to the spot in the room where the evidence is located. Midrange shots are useful for documenting the overall condition and location of evidence at a crime scene, and measurement scales should be used to begin to serve as a reference of size and distance in the photos. When using measurement scales in forensic photography, there should always be at least two scales in each photo, with one scale laying perpendicular to the other. This is necessary to allow the length and width of the evidence to be accounted for in each photograph, as well as to serve as a reference for the angle at which the photo was taken (for more on this, see the "American Board of Forensic Odontology Bite-mark Scale" section above). It is important to note here that midrange shots can be taken at any angle necessary to properly document all aspects of the evidence and conditions at the crime scene. Also, at this point, a forensic photographer should be working side by side with a crime scene investigator in a thorough search and analysis of the scene, documenting any and all evidence that the crime scene search reveals. When necessary, close-ups, or *detail* photographs, should be taken.

Detail photographs at a crime scene are taken to document the fine details of the smaller aspects of the evidence at the crime scene, and to capture these details in a specific way. For detail photography, it is again necessary to utilize the two-dimensional scale approach, but at this point detailed photographs should be taken at a 90-degree angle to the evidence. Also, since these photos should be capturing the fine detail of the evidence, proper lighting is critically important. Depending on the circumstances, it might also be important to photograph the evidence in a *1-to-1* ratio. When taking a 1-to-1 photograph, the focus of the lens is locked so that the image size on the film will be exactly the same size as it is in real life. 1-to-1 photography is commonly used when photographing fingerprints at a crime scene before they are dusted, and again after they are dusted but before they are lifted. The resulting photograph slides can then be directly scanned into the AFIS fingerprint database and their size will be exactly the same size as they occurred in real life.

Forensic Photographic Equipment

To ensure that evidence is documented correctly, a forensic photographer needs the proper equipment. However, the best cameras, film, filter, flashes, and lenses are useless unless the photographer knows how to properly utilize these components to their full advantage. This combination of knowledge and specialization are critical so that the forensic photographer can capture for perpetuity all the necessary details about the crime scene and/or the evidence.

THE BASICS OF FILM AND FILTERS

The basic photographic equipment setup for any good forensic documentation consists of a basic medium-to-large format camera with a variety of lenses, a good flash, backup batteries, a variety of filters, and of course film. The typical film that is used for forensic purposes is standard 35mm film. 35mm film is measured across the width of the entire negative film, so that a basic negative image has a typical 2:3 width-to-length ratio. Film can also be black-and-white or color, and one type of film may be preferred over another sometimes when photographing certain aspects of a scene. There is even a special type of film that is only sensitive to infrared heat, and is used in special circumstances.

Black-and-white film, which is also referred to as *monochrome* film, is very good at documenting *contrast* differences. Contrast can be thought of as the varying degrees of lightness and darkness of an object, surface, or a light source. Contrast photography, in forensic terms, is therefore used to highlight the differences of one substance on another in terms of evidence, such as a dusty footprint on a dark floor, or a greasy fingerprint on a television screen. Filters can also be used to help enhance the differences in contrast, and as a result can have an additive or subtractive effect on the resulting photograph. For example, if a blue filter is placed over a lens and a photograph is taken of a bloody handprint on a blue wall, in the resulting photograph, the wall will be much lighter and the red handprint will be much darker. This is because color filters lighten up their same colors while darkening their *complementary* colors in contrast photography. A complementary color can be thought of as the subtractive color to any other color; when the two are mixed, the result is a darker color that "cancels" out both of the original colors. For example, green and magenta are subtractive colors, as are blue and yellow, and red and cyan. Using this principal in black-and-white contrast photography can help highlight images on the resulting photos by using color filters. Color photography, on the other hand, is used extensively to document the basics and overviews of a scene, but can also be used to show the differences in the contrast of colors of evidence at a scene or in the lab. For example, a high-quality color film with a large degree of contrast helps to show dark-red bloodstains on a black or dark-red background, whereas they would not be clearly seen with the naked eye.

Another important aspect of film is the speed of the film, which is commonly known as the ISO of the film. ISO is a short term for the International Organization for Standardization. ISO is not an acronym, but a modification of the Greek word *isos*, which means "equal." The ISO of film is an international standard that measures the sensitivity at which the film surface collects light. To understand this, it is important first to understand what film is and how film is produced.

Film itself is made up of several components, the main being the actual thin plastic backing that supports the negative. The second component of the film is a gelatin layer, which is adhered to the plastic and holds the final layer in place. The final layer is an emulsion of silver halide salts. The silver halide is light sensitive, and is white until it reacts with photons to turn black. The size of these silver halide salts change from one type of ISO film to another, such that a "slow" film uses smaller grains of salt that require more light and a longer exposure time, and a "fast" film uses larger grains of salt that quickly collect light and require a shorter exposure time. Therefore, the ISO of the film is actually a measure of not only the "speed" of the film but the grain size of the film as well. For this reason, "fast" films tend to have more "grainy" photos, as the result of the large grains of silver halide, whereas slower films tend to display higher levels of detail, clarity, and sharpness (assuming the photo was taken correctly) due to the smaller grain size. This ISO measurement applies equally to both monochrome and color film, and therefore it is important to have an assortment of different types of ISO films on hand when taking forensic photographs.

THE CAMERA

While the speed of the film is important to keep in mind, the amount of available light and the speed of the shutter on the camera are also key elements in capturing a quality photo. While there is a large variety of camera models, styles, and types commercially available, the standard camera that is used in forensic photography is an SLR, or single-lens reflex, camera. Figure 4-2 depicts the standard features of an SLR camera.

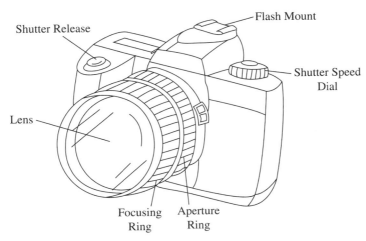

Figure 4-2   The typical features of an SLR style camera

This type of camera directs the image that is seen through the lens to the viewing eyepiece on the top of the camera using a 45-degree mirror that is positioned in front of the camera shutter. This is known as "through-the-lens" (TTL) focusing, because what the photographer sees in the eyepiece is exactly the same image that will be captured on the film. In other types of 35mm cameras, such as the common point-and-shoot varieties, the viewfinder lens is separate from the actual lens of the system. The image that the photographer sees through the eyepiece of these point-and-shoot cameras is not the same image that will be captured on the film, due to the offset of the film lens and the viewfinder lens. Therefore, this style of point-and-shoot camera is not desirable for forensic purposes.

In an SLR camera, as a photograph is taken, the 45-degree mirror is pulled or pushed up and out of the way of the light path so that the entire image is captured on the film when the shutter opens. The speed at which the shutter operates is known as the *shutter speed* and is measured in fractions of a second. The faster the shutter opens and closes, the less light that is transmitted to the film's surface; the slower the shutter operates, the more light that is transmitted. Interestingly, as shutter speed is increased from one setting to the next, the amount of light allowed to pass is exactly halved. But, as the shutter speed is decreased from one setting to the next, twice as much light is allowed. Therefore, it is useful to coordinate the shutter speed with the ISO of the film that is being used to achieve the optimum exposure of the image on the film.

LENSES

Another nice feature about SLR cameras is that most models allow for the quick interchange of different styles and types of lenses. This enables a photographer to use one camera body and interchange several different lenses depending on the necessary shot. Therefore, in forensics, the initial establishing shots can be done with a wide-angle lens, the midrange shots can be done with a medium-format lens, and the detail shots can be done with a close-up, or *macro,* lens. The major difference in these lenses is the *focal length* of the lenses, which is the distance from the lens to a point where parallel light is brought into focus.

The *aperture,* or the opening of the lens, is another key element that can affect the development of photographs. The aperture is actually a circular ring that can be opened or closed to allow more or less light through the lens, similar to the iris of your eye. When the aperture size is correlated against the focal length of the lens, this is known as the *f-stop*. The f-stop on most SLR lenses is expressed as a number, typically as f/#, where f/# is equal to the aperture divided by the focal length. When f/# is small, the aperture is wide and more light is allowed though the lens. When f/# is larger, the aperture is small and less light is allowed through the lens. This affects not only the exposure and development of the image on the film, but also the *depth*

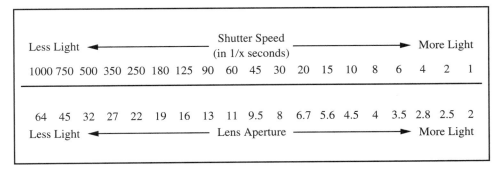

Figure 4-3   Shutter speed vs. aperture scale

of field of the photograph. The depth of field is a measure of what is in focus surrounding the subject being photographed. The larger the depth of field, the more in focus the surrounding objects in the foreground and background will be in a photograph. The smaller the depth of field, the less in focus the surrounding objects will be. This is directly correlated with the aperture of the lens, such that the smallest aperture (and therefore largest f/#) produces the largest depth of field, and the largest aperture (and therefore smallest f/#) produces the smallest depth of field. Aperture is also based on a logarithmic scale, not a linear scale. Therefore, when you move from the smallest f-stop to the next one up, you halve the amount of light that is allowed to pass through the lens. Conversely, if you start at the largest f-stop and move down, you double the amount of light passing through the lens (see Figure 4-3).

In the end, the aperture and the shutter speed have the most critical roles in capturing a quality photograph. Because the change in either direction of either the aperture or the shutter speed can halve or double the amount of light, it is sometimes not clear which setting will be best, and a photographer might choose to *bracket* exposures of one set of shots. Bracketing is a process in which a successive series of photographs is taken but each photo has one change in either the shutter speed or the aperture, but not both. The reason for not changing both is that if the aperture of the lens is moved up one f-stop (to let through one-half of the light) and the shutter speed is decreased (to let twice as much light through as before), the photo would be the same as the previous photo, because the same amount of light overall is still hitting the film. Bracketing only one of the settings enables a photographer to capture the correct photograph in a group of shots.

LIGHTING

Another critical aspect of photography is the proper use of lighting and flashes. Because photography is nothing more than using light to create images on film, it is absolutely necessary for the proper lighting to be available when shooting a photograph. Sometimes, this is achieved by using sunlight and either direct or indirect shading to provide the correct levels of light for the photos. However, for crime scenes that are indoors or photographed at night, a forensic photographer must rely on other light sources, such as lamps, strobes, and flashes, to properly illuminate the evidence that is being photographed. The use of flashes is very common in all types of photography, but in forensics, it is important to keep the flash from interfering with the outcome of the final photograph. Flashes can interfere by either under- or overexposing a photograph if used incorrectly, or can create glare on a surface, which can mask the actual evidence in a photograph. These common lighting problems can be easily avoided with the use of ring flashes and oblique lighting when photographing evidence.

A *ring flash* is a type of flash that attaches to the end of the lens and surrounds the lens. This is useful in midrange and detailed forensic photography, because it limits the distance of the flash to the object being photographed and can cut down on the amount of glare and bounce that is created by normal flashes. Also, if the object that is being photographed is close to the lens, a normal flash might not illuminate the object due to the angle of the normal flash to the object. Using a ring flash therefore creates a more even and flat flash effect, and in effect can produce shadowless figures. This means, from a forensic standpoint, that the image captured on film will be as realistic as possible and not look like an object photographed with a normal flash.

Oblique illumination is not a type of flash, but rather is a special technique that can be used with a standard flash on an SLR camera. Oblique illumination requires that the flash be connected to the camera by a cord or flash trigger so that the flash can be held off to the side of the object being photographed. When the photo is taken, the flash off to the side sends oblique light across the surface of the object rather than directly at it. The benefit of this technique is that the surface of the object "bounces" random photons of the oblique light back toward the lens, allowing the photo to show the contours and details of the object's surface that otherwise would not have been visible. A good example of this can be done by planting a fingerprint on a clear, flat surface, like a CD jewel case or the lens of a pair of glasses. If you then take a flashlight and point it straight at the print, you might have trouble making out the fine details of the ridge patterns. However, if you hold the flashlight at an oblique angle to the printed surface, the light acts as an oblique illuminator and you can see the fingerprint much easier (see Figure 4-4).

One final aspect of flashes and illumination to discuss is a handy technique known as "painting with light." This requires at least two people and a very steady camera tripod. Consider a dark night scene or pitch-black indoor scene that requires establishing shots but has no major light sources, lamps, or strobes available. If a single flash is available, then there is no problem. Simply mount the camera on the

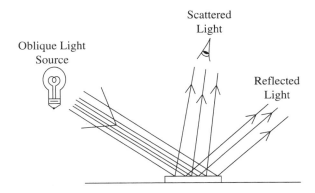

Figure 4-4　Oblique light vs. scattered and reflected light

tripod and set the shutter to remain open until the trigger is pressed a second time, or have the second person hold the shutter open manually. By holding the shutter open indefinitely, the film is exposed to any and all light at the scene, so the lens might need to be temporarily covered with a black card by the second person as well. As the first person mans the camera, another person simply walks around the perimeter of the scene, firing the flash every few feet. As the flash is fired, the lens is uncovered and the flash burst "paints" the scene on the film. As the general area of interest is painted with light, the final image will show what appears to be the scene fully illuminated, with bright flashes in the background where the person working the flash was standing!

WHITE BALANCE AND COLOR TEMPERATURE

On a final note, all photography, forensic included, relies on the proper white balance and color temperature in all of the photographs for accurate and realistic reproduction of true-to-life images. *Color temperature* of light is based on a "theoretical black-box radiator" that gives off a certain color based on the amount of energy, in Kelvin, that is supplied to the box. As an approximation, natural sunlight is around 5500 Kelvin, and most flashes, lights, and other photographic illuminators strive to achieve this same color temperature with their lights. *White balance* in photography is therefore a measure of the color temperature and how it affects the "color" of white. At 5500 Kelvin, white is absolute white, but as the color temperature drops, white can become yellowish-red, and as it increases, white can become blue. Finding this balance and correcting for it can be achieved by photographing an 18 percent grayscale card to calibrate the internal light meter of the camera or an external light meter, if necessary.

Digital Photography and Forensics

The recent advancements of digital cameras, along with their decreased cost, have made many people take notice of the advantages of digital imaging over that of traditional film. In the forensics field, however, there are issues with using digital photography that have both positive and negative effects on the documentation of evidence and crime scenes.

To fully understand the reasons for and against digital photography in forensics, it is important to understand what the differences are between digital and film cameras. In a majority of the digital cameras on the market today, a *charge-coupled device (CCD)* chip has replaced the film, and images are stored on media cards or in the memory of the camera itself. The CCD chip is a photosensitive device that captures and records the image that is focused onto the surface of the chip. The CCD chip is not one device, but instead a series of millions of photosensitive receptors. The overall image quality of a digital photograph is determined by the size and number of receptors on the CCD surface. Because each receptor accounts for a single *pixel,* or dot, of the digital image, the quality of a digital camera is measured in megapixels (1 million pixels). Therefore, the higher the quality of the digital camera, the more megapixels that are available on the CCD. Since there are limitations to the total size of the CCD chip, as the number of megapixels goes up, the relative size of each of the photosensitive cells on the chip gets smaller. To compare this to film, the megapixel count is much like the ISO of the film. The average ISO 1000 film image is therefore around 7–8 megapixels, and most inexpensive digital cameras are still in the 6–7 megapixel range today, while the higher-quality digital cameras are around 8–9 megapixels.

The images that are processed and stored in a digital camera are usually in some common graphics format, such as .jpeg, .tiff, or .bmp, depending on the settings and the quality of the camera. Because the images are stored in memory after they are captured, many digital cameras have a small LCD screen on the rear of the camera that allows the user to view the image after it is taken. This is extremely helpful when trying to capture a difficult image, and the instantaneous feedback can be used to adjust the camera settings and reshoot the image if it is not successful. Also, some of the point-and-shoot digital camera models use the LCD screen as a viewfinder, eliminating the need for the offset viewfinder and allowing the user to "see" through the lens as if the camera were an SLR-style camera. Since the images can be immediately reviewed, the use of digital cameras has all but eliminated the need for bracketing a series of photographs.

The photographs themselves also not only contain the images that were captured by the camera, but most digital cameras also automatically include a data file with each photo. This data might include a time-and-date stamp and readout of the settings that were used on the camera to shoot the image. Some cameras even have indicators of the levels of white balance and contrast as part of the information of the image. Since these images are in a purely digital form, they can be uploaded or downloaded to a computer for viewing, printing, or editing. They can also easily be deleted from the camera, memory card, or computer hard drive.

While all of these issues seem to be benefits that support the use of digital photography in forensic science, several of them are subtle problems that can have massive implications in the proper documentation of evidence. First and foremost, there is a major problem when attempting 1-to-1 photography with a digital camera. A majority of digital SLR cameras use a CCD chip that is smaller than typical 35mm film, and therefore the image after processing is also smaller than the original image when shot at 1-to-1 with a locked lens (see Figure 4-5). For this reason, most cameras have a correction factor that is published in the manual, and this factor needs to be multiplied by the lens lengths to establish what the true 1-to-1 should be. For example, if the correction factor for a CCD is 1.5, a 24mm lens will actually act more like a 36mm lens (24mm × 1.5 = 36mm), and this adjustment needs to be considered when shooting all types of digital photographs. Recently, however, a few high-end, professional SLR digital cameras have come out that have a true 35mm-size CCD that is capable of true 1-to-1 photography, but these are expensive.

The cost of digital photography also needs to be considered when addressing the needs of a forensic department. While the use of digital cameras eliminates the expenses of film and film processing, digital cameras tend to be more expensive than regular film cameras, and this is especially true with the digital SLR models that are on the market. Also, unlike film cameras, digital technology is constantly getting better and better, such that if a forensic lab were to invest in a digital camera system, it would be obsolete in a very short period of time. Therefore, the cost of keeping up to date with the current technology may be prohibitive to many forensic laboratories.

There are also limitations to digital photography that cannot be overcome with technology. Currently, there are very few digital cameras that capture true monochrome (black and white) images. Most digital cameras simply "turn off" the color

1:1 Photo on Film   Original Print   1:1 Photo on Digital

Figure 4-5   1:1 photography on film compared to digital

sensors on the CCD chip, capturing only the contrast differences of an image, but as a result the sensors that are turned off reduce the overall image quality by half or better. Also, there is no digital CCD chip in any current digital camera that captures infrared images, and this is occasionally a useful tool in forensic photography.

Another factor to consider with digital photography is that the images need to be archived and backed up on a computer system. Because digital photo files can be very large (several megabytes), a modern computer system or network with adequate storage capacity is necessary to maintain the proper archiving and storage requirements. This same issue also applies to film photography, but 35mm slides are easily stored in a variety of ways, from books to carousels to canisters, and since processed film is stable when stored under proper conditions, there is no need for redundant backups.

Finally, and perhaps the most controversial issue surrounding digital forensic photography, is the fact that digital images can be easily manipulated. Using readily available programs like Photoshop or other image-editing software, a person can do anything from remove red-eye to add false elements to a digital photo. Likewise, there are specific enhancements that can be done to a digital image to highlight, clarify, or magnify key elements in a photo. Both of these actions have benefits as well as disadvantages in forensics. Obviously, if the use of a software program can help enhance a bloodstain on a dark background, this is helpful. At the same time, the falsification of evidence in a photograph is a major problem.

While there are ways of detecting such manipulations in digital images, the recent advancements in digital editing software can make such changes imperceptible to the untrained eye. As a result, there is a growing concern that an unethical forensic scientist could "digitally" plant a fingerprint or bloodstain in a crime scene photo to bolster the evidence in a case. While there would be dire consequences for any forensic scientist caught doing such an outlandishly unethical act, it does not mean that the concern is unfounded. Currently, the FBI and the American Academy of Forensic Sciences have both established guidelines based on the Scientific Working Group on Imaging Technology's (SWGIT) recommendations for the preservation and documentation of digital evidence. These guidelines form a basic plan that enforces a protocol for digital image authentication, management, and security. In general, the guidelines propose that the original image be preserved, and any manipulation or enhancement that is performed on the image be logged and documented in a forensic manner. In the end, any image that is presented should be annotated as an "enhanced" or "modified" image, and should be shown alongside the original, unaltered image. As digital photography becomes more advanced and the issues surrounding the use of digital cameras are resolved, these guidelines will need to be modified and solidified to ensure that unscrupulous actions are not taken with the images, and that the use of digital photography can be a reliable and valid form of forensic documentation.

Other Types of Forensic Documentation

While photography is the most common form of visual documentation used in forensics, other forms are commonly used as well, and in the end they also serve as a visual medium for collecting and preserving the conditions of the crime scene and the integrity of the evidence. One such form is forensic *videography*, or the use of a video camera to capture moving images over a period of time. While it may seem out of place to consider the use of a video camera or camcorder at a crime scene, it is important to realize that video cameras are commonly used in police cruisers. These cameras not only record the activities of a police officer to ensure that proper procedures are followed, but also record the license plates and vehicle makes and models of the cars that are pulled over. Although this video might not follow the "rules" of proper forensic documentation, it is nonetheless admissible in court and can serve as evidence in an investigation. In the same way, a video camera at a crime scene can be used to document the scene over time as well. The use of a video camera at a crime scene can help in recording the spatial relationships of different areas of the scene, by "walking" the camera through the entire scene. As this is done, the video can provide another overview of the crime scene and begin to highlight key areas of the scene that might be critical to the investigation. Also, just like overview photography, the video can show the changes and interactions at the scene and serve to record the actions of the forensic investigators. It is very important to note that videography should never replace the use of photography at a crime scene; a video camera is simply another tool for general documentation. Videos tend to be grainy and noisy, so it is important that overview, midrange, and detail photography still be done for proper forensic documentation.

Another form of visual documentation can be done with surveying equipment, especially the use of computerized line machines and GPS-linked surveying optics that can map a crime scene. One such machine is designed to capture a 360-degree view of a crime scene, and by plotting key locations, such as building walls, fences, and trees, it can create a three-dimensional computer model of the crime scene. This can be extremely useful when dealing with outdoor crime scenes that cover large areas, and where wide-angle photography alone will not properly account for the establishing shots. Other types of surveying equipment use lasers and distance-determining calculations to form 3-D models of locations or objects at a scene. For example, stairwells, which can be difficult to properly document in photographs, can be mapped in 3-D using the laser rangefinder machine, and a computerized model of the stairwell can be used to map the major locations of the evidence in the case and supplement all the photos of the stairwell as well. As always, as was the case in videography, these surveying tools should never be used in place of photography, but should help by bringing together a crime scene and establish more of an overview than a 2-D photograph can provide.

Forensic Image Analysis

Another important aspect of forensic photography is the analysis of forensic images. There are times that the crime scene or evidence is no longer available to the analyst, and all that is left are the photographs, videotape, notes, and other documentation from the crime scene. If all of the documentation was done correctly, and adheres to the proper forensic protocols, there is no reason that the photos and other items cannot be used to analyze and completely re-create the original crime scene. One of the most common practices in forensic crime scene analysis is the reconstruction of the events of the scene by using blood-spatter or ballistic-tracing analysis. If the original photographs were taken properly, the reconstruction of the trajectory of the blood or bullets can be done to find the origin of the spatter or gunshot. The reconstruction can be accomplished by scanning the original photograph into a computer, or the trajectory lines can be directly traced on a copy of the original photograph. Likewise, the photographs themselves can help an investigator who was not at the actual scene get a first-hand look at the original state of the evidence before it was collected, packaged, and shipped to the laboratory.

A second type of forensic image analysis is an in-depth analysis of the image itself. By using high-resolution scanning and close-up photography on suspected forged or hoaxed images, the modified and edited regions can stand out in a photograph and prove to a trained image examiner that there was some type of manipulation of the original image. Other times, it might be necessary to enhance faded images or re-create missing areas of photographs as well, and there are a variety of software programs that allow for the forensic reconstruction of these images. Because this is considered manipulation and enhancement of digital evidence, the guidelines concerning the proper forensic documentation, preservation, and annotation of the images must be followed.

Forensic image analysis can also be applied to video and other types of motion-picture film. It is a routine practice for automotive safety engineers to use high-speed video to capture the results of crash-test experiments. The results of these videos upon playback allow the engineers to perform a slow-motion or even frame-by-frame analysis of the damage sustained to both the vehicle and the crash-test dummies inside the car.

Sometimes, the videos or films of events that are intended to be nothing more than basic documentation are used later on for forensic purposes. This was readily the case during the investigation in 2003 into the loss of the space shuttle *Columbia*. Cameras on the ground recorded footage of a piece of foam insulation breaking free and impacting the left wing of the shuttle. This impact caused a breach in the heat shield of the orbiter, and as a result, during re-entry into Earth's atmosphere, superheated air seeped into the body of the shuttle, melting the aluminum frame and destroying the shuttle, killing everybody on board. Following this disaster, an extensive review of the camera footage identified not only the exact location from which the insulation broke free, but also the exact location where it struck the body of the shuttle and exactly what heat-protection tile was damaged that led to the fatal final re-entry of the space shuttle.

The same type of image analysis most famously has been applied to the images of the assassination of President John F. Kennedy. The Zapruder film, named after the man who shot the film, Abraham Zapruder, records the final moments in the life of President Kennedy on November 22, 1963. The Zapruder film contains only 27 seconds of 8mm film, but has become one of the most widely analyzed pieces of footage in the world. Despite the grainy nature of the film and the rudimentary style of camera it was shot with, there are very clear moments in the film that show the fatal shots hitting President Kennedy as he rode in his car through Dealey Plaza in Dallas on that day. These individual scenes were scrutinized not only by the experts on the Warren Commission (the unofficial name given to the President's Commission on the Assassination of President Kennedy, based on the name of its chairman, U.S. Chief Justice Earl Warren) in the 1960s, but also by several other analysts over the years that have contributed to the debate over the events and circumstances surrounding the assassination. While the Warren Commission has used the Zapruder footage to bolster the idea that Lee Harvey Oswald acted alone and fired two fatal shots, other experts have used the same footage to show the impossibility of a "single bullet theory" and point out inconsistencies in the actual footage as well. Despite the forensic applications and increases in technology, the analysis is limited by the length and quality of the original footage. As a result, the controversy surrounding the Kennedy assassination and the Zapruder footage will likely continue well into the future.

Quiz

Refer to the text in this chapter if necessary. Answers are located in the back of the book.

1. A wide-angle lens has:
    (a) A narrower angle of view than a telephoto lens
    (b) A longer focal length than a normal lens
    (c) A wider angle of view than a normal lens
    (d) A longer focal length than a telephoto lens

2. Focal length is defined as:

   (a) The equidistant point of focus between the lens and image

   (b) The distance from the lens at a point where parallel light is brought into focus

   (c) The distance from the lens at a point where oblique light is brought into focus

   (d) The distance from the lens to the plane where the object is brought into focus

3. In film photography, the silver halide salts turn black by which method?

   (a) Exposure of the sliver emulsion to photons of light

   (b) Processing the film in developer

   (c) Fixing the latent image in fixer

   (d) Washing the film in acetic acid

4. A telephoto lens has:

   (a) A longer focal length than a normal lens

   (b) A shorter focal length than a wide-angle lens

   (c) A shorter focal length than a normal lens

   (d) A wider angle of view than a normal lens

5. In establishing photographs of a crime scene, what is required to be in the photograph?

   (a) The area of interest around the crime scene

   (b) The areas where detailed photos will be taken later on

   (c) A landmark

   (d) All of the above

6. What color of color contrast filter would be used to enhance a bloody handprint on a blue wall?

   (a) Red

   (b) Blue

   (c) Green

   (d) Orange

7. In digital cameras, the acronym in "CCD chip" stands for:

    (a) Closed-circuit diode

    (b) Camera-card device

    (c) Charge-coupled device

    (d) Current-carrying digit

8. To cut down on the amount of glare and bounce from a normal flash, one could use:

    (a) An oblique flash

    (b) A ring flash

    (c) Natural sunlight

    (d) All of the above

9. SLR stands for:

    (a) Single-lens reflex

    (b) Standard-lens recoil

    (c) Single-laser reflection

    (d) Side lacking ridges

10. Most digital cameras can mimic every aspect of film photography except:

    (a) Zoom photography

    (b) Wide-angle photography

    (c) Flash photography

    (d) Infrared photography

CHAPTER 5

Forensic Pathology

In the lobby of the Chief Medical Examiner's office in New York City, there is a phrase in Latin, on the main wall behind the security desk:

Taceant colloquia. Effugiat risus.
Hic locus est ubi mors gaudet succurrere vitae.

The phrase is best translated as "Let conversation cease. Let laughter flee. This is the place where the dead come to aid the living." The Office of the Chief Medical Examiner in New York City is where autopsies are performed to investigate deaths that occur within the city. Most major cities have a building like this one, or use city hospitals to house the forensic death investigation services. Regardless of where the facility is located, the phrase above is an important one, and clearly states the purpose of the building and the work that goes on inside it. The primary mission of the medical examiner (ME) is to investigate cases of persons who die from criminal violence; by casualty or by suicide; suddenly, when in apparent good health; when unattended by a physician; in a correctional facility; or in any suspicious or unusual manner.

Forensic pathology is the medical and legal process of investigating the cause and manner of death. Pathology, in normal medical terms, is the science of disease and dysfunction. If you have a biopsy done on a possible tumor, the sample is sent to the pathology department in the hospital, where the doctors perform tests to see if it is benign (not harmful) or malignant (a rapid-growth tumor). In forensic science, pathology still deals with disease and dysfunction, but usually only in the process of investigating a death.

Forensic pathology has several aspects that play a role in the investigation process. First and foremost is the *autopsy*, a process in which the body is examined externally and internally by a forensic pathologist to find the cause and manner of death. During an autopsy, samples of internal organs, body fluids, and foreign objects may be recovered by the forensic pathologist, and these are usually sent to other labs within the ME's office for further testing. If, for example, a disease is suspected, the tissue samples can be sent for histological testing.

Histology is the science of cells and tissues. A histology lab prepares a tissue sample on a microscope slide and "stains" the tissue to highlight certain diseases, abnormalities, or functions of a cell that can give a forensic pathologist clues to how somebody died.

Likewise, body fluids and other foreign objects can be sent for toxicological testing. *Toxicology* is the science of toxins and poisons. A toxicology lab can screen body fluids and tissue samples for drugs, toxins, or poisons. If there are drugs, toxins, or poisons present, the toxicology lab can determine how much of the substance is present in the body, and if it was in fact at lethal levels. All of this information serves to aid the forensic pathologist in his or her job of answering the questions surrounding a suspicious death, discussed next.

Who, What, When, Where, and Why?

These are the basic questions of any investigation of a suspicious death. The forensic pathologist's job is not to answer these questions themselves. The true job of the forensic pathologist is to interpret what the body reveals, and to be a voice for the dead!

QUESTION 1: WHO IS THIS?

Identification is a major issue in any forensic investigation, and this is true as well in forensic pathology. Identification can be done in several different ways. A body may be visually identified by a relative or loved one, or can be accomplished simply by checking the photo ID that is found on the deceased. Sometimes the identification cannot be done based on facial recognition; scars, marks, tattoos, or piercings that are unique to the individual must be used instead.

Many times, a body is found without ID or is so badly decomposed that it cannot be recognized at all. In such an instance, *ante-mortem* (prior to death) dental records, when compared by a *forensic odontologist* to the *post-mortem* (after death) dental x-rays of the unknown body, can help make an identification. The use of ante-mortem x-rays can also highlight old broken bones, surgical implants, or old wounds, and can be matched against corresponding bones, implants, or wounds on an unknown body. There even have been instances in which unknown bodies have been identified by matching the serial numbers found on artificial hips, breast implants, and pacemakers to medical records of the deceased owners.

If none of these methods produces an identification of an unknown individual, a forensic pathologist might then rely on fingerprinting or DNA analysis to help provide a name. The importance of identification of the body is not only to establish the name and identity of the deceased, but also to be able to notify the family of the deceased so that they will have the opportunity to oversee the internment of the remains.

QUESTION 2: WHAT KILLED THIS PERSON/HOW DID THE PERSON DIE?

When a forensic pathologist begins investigating a death, there is one main question specifically about the death that must be addressed: what was the *manner* of death? The manner of death can be only one of five possibilities—*homicide, suicide, natural, accidental,* or *unknown*. It is up to the forensic pathologist to take into account all of the information surrounding the case, including the findings from the autopsy, reports from histology and toxicology, and even aspects from the police investigation and crime scene details, to make a final decision on the manner of death.

The manner of death is the most important finding in a death investigation case, and can be made only by the forensic pathologist. For example, if there is a body found at a crime scene and the police have a suspect in custody, if the manner of death is found to be a homicide, then the suspect can be formally charged with the victim's death. But, on the other hand, if the forensic pathologist finds that the victim died from natural causes or was a suicide, then the suspect is set free. The only problem with establishing the manner of death is the fact that one of the categories is "unknown." A forensic pathologist may opt to report this finding if the evidence from the autopsy, reports, and other information does not clearly answer how a person died, or if there is an equal possibility of more than one manner of death.

QUESTION 3: WHEN DID THE DEATH OCCUR?

The time of death is an important fact that a forensic pathologist must determine in the course of a death investigation. Sometimes, the time of death is clearly known, perhaps by people at the scene of the death, or by medical personnel that were unable to save the person's life. However, if a body is discovered without any clear way to identify the time of death, the forensic pathologist has several avenues of investigation to approximate when the person died.

The first thing a forensic pathologist looks at when determining the time of death is the stiffening of the body, which is known as *rigor mortis*. Rigor mortis is a process in which the muscles of the body contract, because after death there is no chemical energy for muscles to use. As the body begins to decompose, the muscle tissue begins releasing enzymes that break the muscle tissue down and cause it to lose the stiffness. The time in which this happens is fairly constant and predictable, and if a body is kept under careful observation, the stiffening/relaxing of the body can point back to the original time of death.

Rigor mortis usually starts to set in 12 hours after death, at which point it begins to slowly stiffen the outermost extremities of the body, the fingers and toes. Over the next 12 hours, the entire body becomes completely rigid, so that 24 hours after death, the body is rock hard and the arms, legs, and head can be moved only with great difficulty. After the body is in full rigor, it takes another 12 hours for the muscles to relax and the body to return to a limp state. So, if a body is found within the first 36 hours after death, a careful analysis and observation of the rigor can be back-calculated to an approximate time of death. For example, if a body is found in partial rigor (the fingers and toes are immobile and the arms and legs are stiff), and under observation full rigor sets in 4 hours later, and another 12 hours later the body is limp, it can be estimated that the body was found approximately 20 hours after death.

Another important indicator that a forensic pathologist can use to determine the amount of time since death is the body's temperature, or *algor mortis*. Algor mortis is the constant rate at which a dead body loses heat, and in a normal, room-temperature environment is about 1.5 degrees Fahrenheit every hour. Because the average body temperature is 98.6 degrees, the forensic pathologist can use a thermometer to measure the rectal or internal liver temperature of a body. If, for example, the internal temperature of a body is around 95.5 degrees, and the body was found in a normal room, the algor mortis determination shows the time of death to be 2 hours ago (98.5 − 95.5 = 3 degrees; if the body loses 1.5 degrees every hour, then it takes 2 hours to lose 3 degrees). Algor mortis can be tricky, though, because the ambient temperature can speed up or slow down the rate at which a body cools. Therefore, a body found in a freezing meat locker or kept outside on a hot July day will not accurately reflect the real algor mortis rate. A forensic pathologist must take the environmental conditions into account when looking at the body's temperature.

While rigor mortis and algor mortis are helpful indicators for a forensic pathologist, they are only useful within the first two days after death. If a body is found a week or even a month after death, neither rigor nor algor mortis will be useful for the forensic pathologist. Another discipline that can be used to determine the time since death has nothing to do with what's in the body, but rather what's on it. That means maggots, flies, and beetles! Insects have an amazing ability to find dead bodies, and the life cycles of several insects are well known and specific. A forensic pathologist might seek out the expertise of a *forensic entomologist*, also known as a forensic insect expert, to identify the species of insect and to determine the length of its gestational cycle. This can help in giving a better idea of the number of days since death. While insect activity is not as good as rigor mortis or algor mortis in determining the time of death to the hour, it can help in estimating the number of days or weeks since death.

QUESTION 4: WHERE DID THE DEATH OCCUR?

The scene of death is important. It might contain further evidence that can be used in determining the manner of death. As a result, bodies are sometimes moved by the perpetrator to cover up the original, or primary, crime scene. The forensic pathologist, though, can look at key physical characteristics on the body to help show whether or not the body was moved, and if so, where the original scene of death might be.

When a person dies, the heart stops pumping blood throughout the body. As the body's cells begin to break down, the capillaries around these cells burst, causing blood to pool in the tissues of the body. Blood, like all liquid, settles to the lowest spot because of gravity. If a person dies on their back, their blood pools in the backs of their legs, buttocks, shoulders, arms, and head. As the blood pools in these areas, it can be seen through the layers of skin as red splotches. This is known as *livor mortis*. These splotches begin to form immediately after death, and after a certain amount of time become permanent, or *fixed*. While this can aid a forensic pathologist in determining the time since death (see the previous section), it is most useful in showing that a body was in the same position when the person died. For example, if a person died on his back and livor mortis set in, and then the body was moved and placed face down, the livor mortis would still show up on the back of the arms, legs, shoulders, head, and buttocks. This would prove that the body was moved post mortem. Moreover, livor mortis cannot occur where the tissues are being squeezed or pressed. For example, if a person died sitting in a chair, the blood cannot easily pool in the back of the legs or on the buttocks because the weight of the person is compressing those tissues. Livor mortis would be present on the lower back and on the feet as well. If the body was moved from the chair after the livor was fixed, the backs of the thighs would show *blanched* areas where no blood had pooled while the surrounding tissues would be a deep red.

While livor mortis can help show that a body was moved, it does little to help locate the original crime scene. To do this, a forensic pathologist must carefully collect the *trace evidence* from the body and rely on trace evidence experts to analyze whatever was found on the body. It is therefore important that forensic pathologists be well skilled in all aspects of forensics, not just pathology, so that they can find all of the evidence that the body has to reveal.

QUESTION 5: WHY DID THE DEATH OCCUR?

The question of "why" is just as important as the question of "what" in a death investigation. The "what killed this person?" question led to the manner of death, but the question of "why did this person die?" deals with the *mode*, or cause of death. The mode of death is what led to the manner of death, and can be any number of reasons. For example, the mode of death in a homicide could be anything from a shooting to a stabbing to a poisoning, or a combination of all three. If you think about it, though, these same modes could also be the reasons for a suicide or an accidental death.

Sometimes, the actions that lead to a death can be the same from case to case, but the effects of the actions result in different modes of death. For example, in a homicide, a shooting might have ruptured a lung, leaking air into the chest cavity and causing the lung to collapse and the person to ultimately suffocate. (This is known medically as a *pneumothorax*.) Therefore, the autopsy report would indicate that a shooting (mode 1) led to a pneumothorax (mode 2), which led to suffocation (mode 3), which resulted in a homicide (the manner of death). Alternatively, sometimes a shooting is not instantly fatal, but the projectile is retained in the body. If the bullet later dislodges, it can work its way into the circulatory system and get lodged in the brain or the lungs, cutting off the blood supply to these organs, causing death. This could be months to years after the shooting, but since the bullet from the shooting (mode 1) led to the fatal blockage (mode 2), which led to the death of the person, the death is still ruled a homicide (the manner of death). Determining the mode of death is therefore the second factor in death investigation that is the singular responsibility of the forensic pathologist. Although the facts of the investigation are gathered by all of the investigators involved in the case, the forensic pathologist ultimately has the responsibility to issue the final death certificate that outlines the mode and manner death.

Who Is Involved in a Death Investigation?

Up to this point, the focus of the forensic pathology discussion has been on the forensic pathologist, but other people may be involved in a death investigation. Sometimes, the local laws and ordinances dictate what type of death investigator is in charge of a case. In some states and counties, the *medical examiner (ME)*, a medical doctor who works at the medical examiner's office, is given the authority. In other areas, a *coroner* is employed as the lead death investigator. Whereas an ME is always a medical doctor, a coroner needs only basic training in determining whether a body is dead, and issues reports based on the autopsy, toxicology findings, etc. that are performed by other specialists. Coroners are also typically elected positions in local or state governments, whereas an ME is an appointed physician outside of the political arena. Although a coroner could be a medical doctor, they typically are not, and therefore rely on a doctor or pathologist for the autopsy report.

Another major difference between a coroner and an ME is that the coroner typically responds to the scene of death or goes to where the body is discovered. An ME might go to scenes, but in specific areas around the United States, medical examiner systems rely on *medicolegal investigators (MLIs)* to attend the crime scenes, collect evidence, and pronounce the body dead. An MLI typically has a degree from a physician's assistant program, and a strong background in forensics. Using such a system, the MLIs attend to the scene of death, allowing the MEs to devote their time to performing autopsies and reviewing the histology and toxicology results. In the end, this allows the city or county that utilizes such a system to investigate a higher number of cases, which is beneficial for the society that is served by the death investigators.

Not Just a Death Investigation

While investigating homicide or unnatural deaths is the mainstay of the forensic pathologist, all deaths, including natural deaths, must be investigated as well. The public health aspect of death investigation is an important secondary aspect to the job of a forensic pathologist. The screening of natural deaths not only is important for the maintenance of public health, but also helps to rule out any and all foul play to ensure that no crime is overlooked.

Not all deaths are criminal in nature, of course, but a single natural death from a contagious disease can have far-reaching implications for the population of a city, state, or nation. There are lethal diseases in nature such as meningitis, smallpox, typhoid fever, and, most recently, avian flu that have the ability to kill or incapacitate hundreds or thousands of people. Therefore, the public relies on the forensic pathologist to identify these diseases before they can become epidemics. As a public health doctor, a forensic pathologist is constantly on the lookout for contagious diseases that lead to death. Because pathologists are usually the first witnesses to an outbreak of a contagious disease, it is up to them to recognize, diagnose, and warn the public of the risks of such diseases.

The Autopsy

When a body arrives at a morgue, hospital, or ME's office for an autopsy, there are several things that should already be known about the body. These are the facts from the scene of death or where the body was found, and should include the condition of the body, the sex of the person (if it can be determined), the ambient temperature and the environmental conditions surrounding the body, and a detailed account of the debris, personal effects, or other possible evidence surrounding the body. If a homicide or a struggle is suspected, the hands of the body should be covered in clean paper bags and sealed with ties to prevent any trace evidence from getting dislodged and lost in the body bag. If the body is coming from a prison or hospital, all of the medical devices or personal effects should remain attached to the body so that the forensic pathologist can determine if the presence of the items led to the death of the person.

As bodies arrive at the morgue, they are usually kept in large refrigerated rooms or coolers to slow down the decomposition process. Just before the autopsy, the body is undressed, weighed, and measured. If necessary, the body is x-rayed and photographed, both to document the condition of the body and for identification purposes.

During the autopsy, several people may be present in the autopsy suite. These include the forensic pathologist, who performs the majority of the autopsy, and possibly other pathologists who are present to observe the autopsy or aid the lead pathologist as *prosectors* (pathologists who dissect the internal organs, looking for signs of disease). Another person who is commonly present is a mortuary technician, whose job is to prepare the body for autopsy by washing it after the external examination, and to assist the pathologist throughout the course of the autopsy with moving, turning, and repositioning the body. At the end of the autopsy, the mortuary technician replaces the removed organs and sews the body up for removal to a funeral home or city cemetery. Additional people who might be present at the autopsy are the police detectives or district attorneys if the case is a suspected homicide, as well as a forensic photographer who is present to document signs of trauma or abnormal findings, or just to record the process of the autopsy.

When the autopsy begins, the body is transferred from the body bag to the autopsy table. This may be a normal, operating room–style table, or it may be a specialized table that includes drains and running water to help siphon off the fluids from the body. Surrounding this table are the various implements that the forensic pathologist uses to examine the body. First and foremost, the forensic pathologist must have a way to document all the findings during the autopsy, whether this is with pen and paper, a wax pencil and a dry-erase board, or a wireless voice-activated dictation recording device. Whatever the medium, all of the information gathered in the course of the autopsy must be recorded, from the weight of the internal organs to unusual smells that the forensic pathologist might encounter during the procedure.

Other implements in the autopsy suite include a scale to weigh the organs, a cutting board for dissection of the organs, jars and buckets for tissue samples, and a variety of scalpels, knives, hooks, syringes, and other medical implements designed to enable the forensic pathologist to find out the body's answers to the who, what, when, where, and why questions surrounding that person's death.

After the body is placed on the autopsy table, the first step in the procedure is the external examination. This is done by the forensic pathologist to look for any outward signs of trauma or unusual features, as well as to look for any outward signs of disease and to get an idea of the general health of the person at the time they died. Also at this time, trace evidence that is on the outside of the body can be identified and collected, including tape lifts from the skin, pubic hair and head hair combings, and, where applicable, scrapings from under the fingernails to see if any of the attacker's skin or other telltale debris might be present. If a sexual assault is suspected to have occurred in the course of the victim's death, a post-mortem sexual assault evidence kit is used to collect possible semen or saliva evidence from the vagina or penis, anus, mouth, or even the skin of the victim. All of this trace evidence is collected and sealed to be sent to other laboratories for testing. If the body is dirty or covered in debris, it is then washed and re-examined to see if any more information is available on the surface of the body. Once all of the external evidence is collected and documented, the pathologist is ready to begin with the internal examination of the person.

To begin the internal examination, the forensic pathologist performs a *Y incision*, which consists of cutting the skin on the chest from shoulder to shoulder across the collarbones, and then cutting from the top of the breastbone to just above the pubic region, which results in a large "Y" carved into the body (see Figure 5-1).

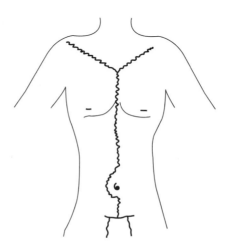

Figure 5-1   Y incision

The skin is then pulled back to expose the internal organs and the ribcage. The breastplate must be removed so that the pathologist can access the heart and lungs, so the ribs are cut using a saw or a device that looks like (and sometimes is!) a tree pruner. Once the breastplate is removed, the forensic pathologist performs a quick overview to make sure that all of the organs are in the proper position and looks for obvious signs of disease (see Figure 5-2). If the death is a violent crime like a stabbing or shooting, the first examination to be done is to find the path of the blade or the bullet in order to see what internal organ was damaged, and if that damage is indeed the primary cause of death.

After the chest and abdomen are opened and the overview examination is complete, the forensic pathologist begins removing the various internal organs, weighing each one, and dissecting each looking for flaws or signs of disease. For example, when the heart is removed it is weighed. If it is within a certain weight range for the age, height, and body weight of the individual, then the heart is considered normal. If the heart is too large, it could be a sign of heart disease, hypertension, or other coronary factors that led to the death. As the heart is dissected by the pathologist or one of the prosectors, the valves are checked for disorder and the coronary arteries are checked for blockages and other signs of heart disease. The same is done with the other organs as well, including the liver, lungs, kidneys, spleen, gallbladder, and stomach. The stomach is unique, in that the contents can be recovered and saved as well. The stomach contents are useful not only as evidence, but sometimes also to help establish the time of death based on known rates of digestion of certain foods. In cases of drug overdose,

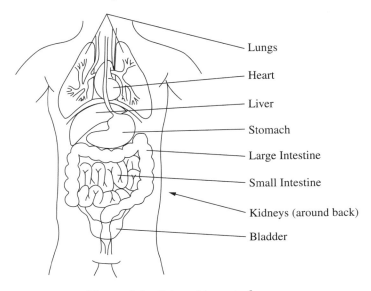

Figure 5-2   Internal layout of organs

the stomach contents can also help identify what kind of drug the decedent ingested based on partially digested capsules or tablets. Finally, the intestines and bladder are removed and drained to check for signs of disease and problems. While this is being done, each of the major organs are sectioned and sampled for histology and toxicology. As the organs are dissected, a small piece of each organ is cut off and placed in a jar with a preservative chemical, usually formalin or formaldehyde. Once all of the major organs have been removed and checked, they are placed in a plastic bag and the bag is returned to the chest cavity, along with the breastplate.

At this point, the major examination on the body is complete, but the forensic pathologist might need to do further dissection on other areas of the body. For instance, if evidence suggests the body was hit by a vehicle, the forensic pathologist may need to dissect the knees and legs to look for signs of fracture or ruptured blood vessels. In another instance, if strangulation or suffocation is suspected, the forensic pathologist may need to dissect the neck to check for damage to the windpipe or hyoid bone (a broken hyoid bone is a strong indicator that the person was strangled). If carbon monoxide poisoning is suspected, the vitreous fluid (the liquid inside the eyeball) can be removed and sent to toxicology. Once all of these secondary examinations are complete, it is time to move on to the head and brain.

To remove the brain, the forensic pathologist or mortuary technician begins by making an incision from ear to ear on the back of the head. This allows the scalp to be pulled forward to expose the skull (see Figure 5-3). This is done mainly to leave the face and forehead intact in case the body will be viewed at a funeral. Seeing this procedure done is by far one of the most unnerving aspects of the autopsy, because the skin is pushed forward as the scalp is reflected, causing it to look like a rubber mask drooping off of the face.

Once the skull is exposed, a special bone saw called a *Stryker saw* is used to cut the skullcap. This saw is a special vibrating saw that cuts only hard objects. Because it's not a spinning saw, it will not damage soft tissue. Once the skull is cut, it is pried off using a tool like a crowbar, exposing the brain. The brain is covered by a thin membrane called the *meninges* (the disease in which this membrane becomes infected and inflamed, causing pressure on the brain, is meningitis), and this is examined and cut open to expose the brain tissue. The brain is partially removed from the skull, and the forensic pathologist then must cut the spinal column to separate the brain from the brain stem, after which the brain is removed from the head. Unfortunately, the brain cannot be immediately dissected at this point. The brain is made of extremely soft tissue, the consistency of which is like runny Jell-O. Because of this, the brain cannot be immediately sliced. After the brain is removed from the body, it is placed in a bucket of formalin for two weeks. Much in the same way boiling water firms up a runny egg, the chemicals firm up the brain tissue,

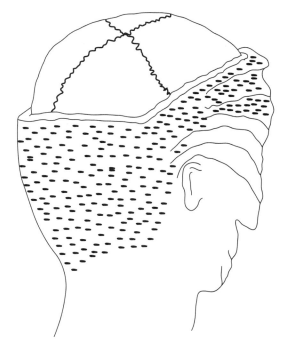

Figure 5-3   Rear view of reflected scalp

allowing it to be easily sliced and dissected for histology, toxicology, and *neuropathology* (the study of the diseases of the brain and nervous system). After the brain is removed from the body, the skullcap is replaced and the scalp is moved back to its original position.

To finalize the autopsy, the incision on the back of the head and the Y incision are stitched up by the mortuary technician, and all of the vital information is transcribed on the death certificate. The body will typically be wrapped in a sterile sheet or be replaced in a body bag if there is no further need for the body, and then sent to a funeral home. The tissue samples are sent to toxicology and histology, and the forensic pathologist awaits the results from those labs while beginning to write up the autopsy notes in a case file. Along with the autopsy information, the file will contain photos from the autopsy, x-rays, documents about the personal effects found on the body, and, if the body was identified by a family member or loved one, a letter of identification. While the mode and manner of death might be known after the autopsy, all of the results from the other labs must be checked to confirm this finding. If after the autopsy the mode and/or manner of death is still unclear, the forensic pathologist must rely on the lab tests to aid in a final diagnosis of the cause of death.

Wound Patterns

As the preceding discussion of the autopsy illustrates, even a simple autopsy takes time and is a complicated procedure, requiring the work of several specialists to uncover the ultimate cause of death. In more complex cases, where there is a violent crime such as a shooting, stabbing, or beating that led to the death, the discovery and documentation of the wounds is critical to the case. Documentation of the wounds not only proves the cause of death, but must also be presented in a court of law as evidence of the criminal act of homicide. There are several different types of wound patterns that a forensic pathologist might encounter in the autopsy of a violent death.

DEFENSIVE WOUNDS

Defensive wounds are typically found on victims who were killed in close proximity to their attacker. If the attacker was wielding a weapon, such as a knife, the defensive wounds usually are found on the hands and forearms of the victim. Because weapons like blades, clubs, and axes require the attacker to be in close proximity to the victim, there might be a hand-to-hand struggle. If this happens, the victim, while attempting to defend themselves, will be injured by the weapon. In a knife attack, it is common to see defensive wounds on the hands or forearms in the form of deep lacerations in the skin. With a club or another type of crushing weapon, the bones of the hands or forearms might be broken or bruised as a result of the struggle. These wounds, when present, indicate a violent struggle and can aid the pathologist in understanding the method and weapon used in the attack. The presence of such wounds also can indicate that the death was not suicide or accidental if the manner of death is not immediately clear.

PUNCTURE WOUNDS

When a person is stabbed with a sharp object, the wound is known as a puncture wound. This wound type can occur with anything from a syringe needle to an ice pick to a blade. Each of these weapons produces a different size puncture wound, which penetrates the body at different depths and causes different injuries. Commonly, the injury from a needle is not lethal but the contents of the syringe might be toxic or poisonous to the individual being stuck. An ice pick or a blade, on the other hand, can penetrate deeper into the body and rupture internal organs or blood vessels, leading to death. During an autopsy, the forensic pathologist must look for these wounds and trace the path of such punctures through the body's layers and organs. Sometimes, a person might be stabbed multiple times in an attack, but only one of the punctures is lethal. All of the injuries must be documented and investigated so that the forensic pathologist can uncover the fatal wound.

LACERATIONS

A laceration is a cut or a slice in the skin, usually made by a blade or a sharp object that is swung or sliced. Compared to a puncture wound, a laceration is wider but not as deep, but can still produce fatal results if major blood vessels are severed in the injury. Laceration wounds usually have smooth edges, depending on the type of weapon or the speed at which the weapon is moving when it contacts the skin. In close-contact knife attacks in which the victim has defense wounds, these wounds are commonly lacerations on the hands and forearms.

ABRASIONS AND CONTUSIONS

Abrasions cover a wide variety of injuries that can occur on the surface or under the surface of the skin. Abrasions can be found in injuries like bruises and scrapes, to entire regions of skin worn away by friction. The term "road rash" is commonly used to describe the effect on the skin of a person involved in a vehicular accident when sliding across pavement. Contusions are injuries that extend deeper into the body, and are commonly the result of *blunt force trauma*, which happens when a body is beaten or otherwise subjected to hard, sudden impacting forces. While abrasions and contusions may or may not be fatal, the discovery of these injuries can provide important clues to the nature of the death, and can therefore aid the forensic pathologist in making a final determination as to the manner of death. For example, a body that has no serious outward injuries such as punctures or lacerations might be heavily bruised instead. Such bruising could lead the forensic pathologist to look for a fatal blood clot that could have traveled to the heart, lungs, or brain and led to the death of the individual.

TEARS

Tearing is a more general term used to describe wounds caused by shearing forces. When the body is subjected to shearing forces, the skin and tissues shred, producing rough, ragged edges and typically large blood loss. These wounds can result from mechanical injuries, but are also seen in violent animal attacks. While lacerations and abrasions tend to affect only the skin and rarely affect the underlying tissues, tears tend to shred all of the underlying muscles, as well as the bones and connective tissues.

GUNSHOT WOUNDS

Gunshot wounds are seen in several different manners of death, from homicides to suicides to accidents. A bullet does damage to a body mainly in two ways:

- The bullet penetrates the body, puncturing the skin and underlying organs, causing rupturing of such organs and blood vessels.
- The kinetic energy of the bullet is transferred to the penetrated skin and organs, causing tears and other expanding-force trauma. Imagine that the kinetic energy is like a balloon that is inserted uninflated into the body. As the balloon is inflated and expands, the edges of the balloon push organs and skin out of their natural positions, causing the organs and skin to rip and tear to allow for the expanding balloon. This type of energy-transfer damage is usually seen only in wounds produced by high-powered rifle rounds or large-caliber pistol rounds.

Because shooting deaths may include more than one wound in the body, it is the responsibility of the forensic pathologist to trace the route of every bullet through the body, and attempt to recover the round if is still present in the body, for comparative-testing purposes.

Another important aspect of gunshot wounds is the distance of the body from the weapon. The closer the victim is to the gun, the more likely two types of gun-shot wounds will be present. The first type includes the presence of *punctuate abrasions* on the skin surrounding the gunshot wound. These wounds are known as *stippling*. Stippling is produced by unburned or still-burning gunpowder moving at high velocity as it ejects out of the barrel following the bullet. The closer the person is to the weapon, the tighter and more pronounced the stippling is on the skin. For the forensic pathologist, the presence of stippling is a sign that the gun-shot was a close-contact shot, and can establish how far away the shooter was from the victim.

The second type of gunshot wound that can help establish the distance of the weapon from the body results from the expansion of gases that follow the bullet out of the barrel (along with the gunpowder). When a bullet is fired, a chemical reaction of the gunpowder causes the powder to ignite and rapidly burn, producing rapidly expanding gases that propel the bullet from behind and push the bullet down the barrel of the gun. These gases follow the bullet out of the barrel, but quickly dissipate into the atmosphere after a few feet. However, if a body is close enough to a gun while these gases are rushing out of the barrel, the gases can enter the wound following the bullet, further expanding the skin and organs. This gas causes not only ripping and expansion-type injuries, but thermal injuries as well due to the heat produced in the burning of the gunpowder. Again, for a forensic pathologist, the presence of such burned, expansion-type wounds surrounding a gun-shot can also help establish how close the victim was to the weapon when it was fired.

Anthropology

If a body is left undiscovered for a long time, when it is finally discovered, the soft tissue and organs may no longer be present. All that remains is the skeleton of the person. In such cases, a normal autopsy is impossible, because clues as to how the person died that would normally be found in wounds in the skin or diseases in the organs have long since vanished. As a result, the forensic pathologist might turn to a *forensic anthropologist* to examine the bones and attempt to answer the questions of who, what, when, where, and why. *Forensic anthropology* is the study of bones and skeletal remains, which is quite different from the classical view of anthropology, which is a social science that studies the culture of large groups or societies. Forensic anthropology has nothing to do with the study of culture or society, and instead focuses on using the bones or other remains of a person to estimate the age, weight, sex, height, race, and even the manner of death. Forensic anthropology is capable of doing this based on key features and differences that can be seen in the bones from one person to another. For example, to establish the sex of a person, the forensic anthropologist might examine only the pelvis. On a male, the pelvis tends to be narrow and flat, whereas in a female the pelvis is wider and more bowl-shaped. These differences are based on the physical uses of the pelvis by the different sexes, and are clear indicators to a forensic anthropologist of the sex of the unknown individual. Other vital information that the forensic anthropologist can find from the bones is the age, weight, height, and race of the person, all by comparing known standards to measurements from the skeletal remains.

Beyond the physical attributes of the body, the bones can also tell the forensic anthropologist how the person died. In violent deaths, bones may be nicked or cut with knives, blades, or axes. Bones may be shattered or crushed by blunt-force trauma from a beating or high-speed collision. Bones may even show the path of a bullet through the body if the bones were hit by the projectile. Bones can even trap bullets, and there has been more than one case of a bullet found embedded in a rib or skull that could be traced back to the original weapon.

Ultimately, bones can be used to identify a missing individual by comparing dental records or known x-rays to the unknown skeletal remains. The condition of the bones can even provide clues to a forensic anthropologist about how long the person has been dead. While time-of-death estimates based on skeletal remains are not accurate down to the minute, a forensic anthropologist can usually narrow down from years to months the time since death based on the overall condition of the bones. Although bones do not offer the same evidence that a whole body offers for a death investigation, they still can provide clues to the forensic anthropologist and pathologist to help them solve the case.

Mass Disasters and Forensic Pathology

In recent years, there have been several notable mass disasters that have claimed numerous human lives, such as the flooding of New Orleans in 2006 after Hurricane Katrina, the tsunami disaster in the Indian Ocean in 2005, and the attack on and collapse of the World Trade Center on September 11, 2001. Various other situations, from plane crashes to destructive earthquakes, can also cause a large number of dead that need to be recovered and examined by a forensic pathologist. When a mass disaster occurs, the death toll could be several hundred to several thousand lost lives, with each of those bodies requiring identification and a ruling on the mode and manner of death.

The challenge of forensic pathology when faced with a mass disaster is the fact that every one of the victims in the disaster requires an investigation into their death. In an earthquake or plane crash, the resulting deaths are usually ruled as accidental. In the World Trade Center disaster, however, the attack was a terrorist act. Therefore, the dead in the World Trade Center disaster were all ruled as homicide victims.

Because the manner of death for the World Trade Center victims was established as a homicide, it made the recovery and examination of the bodies that much more important. As part of the investigation of the bodies, it was determined that the mode of death for most, if not all, of the victims of the World Trade Center was blunt force trauma. This finding was based on the direct examination of the victims' remains. Because of the nature of collapsing buildings, the crushing and tearing forces that took the victims' lives made any further examinations or findings impossible. Once the mode and manner of the deaths were established by the forensic pathologists, the second, and most important, aspect of the investigation began— the identification process.

Identifying the dead is always a goal in a forensic pathology examination. For intact bodies, the issue of identification in a disaster recovery effort is usually done by searching the body for ID, or using fingerprints or dental records to find the name of the individual. In the case of the World Trade Center disaster, there were other factors that unfortunately left the bodies shredded, burnt, and in many cases indistinguishable as human. The identification of body parts, then, became a two-prong approach to the recovery effort. The first step was to identify the body part as belonging to an individual, either through anthropology, odontology, fingerprinting analysis, or DNA. The second step was to match all of the pieces of the bodies back together, again using the same technologies, so that the families could bury all of the remains of their deceased. Because of the scale and scope of the World Trade Center disaster, there were many bodies and body parts that were not identified, and many families left without the closure of being able to bury their loved ones.

In the end, out of the 2749 victims of the World Trade Center attack, the recovery effort was able to identify 1592 of the bodies. Out of those victims, every body, and every body part, was examined by a forensic pathologist and tested to the full extent of what forensic technology would allow, helping bring closure to one of the largest attacks on U.S. soil.

Quiz

Refer to the text in this chapter if necessary. Answers are located in the back of the book.

1. A deceased body at a crime scene is found in full rigor. Under observation, the extremities begin to become flexible after 3½ hours. What is the probable time of death?

    (a) The week before the body was found

    (b) The day before the body was found

    (c) Approximately 20 hours before the body was found

    (d) Approximately 8 hours before the body was found

2. Vitreous humor is found in:

    (a) The brain

    (b) The liver

    (c) The gallbladder

    (d) The eye

3. What body fluid is preferred for carbon monoxide testing?

    (a) Urine

    (b) Cerebral-spinal fluid

    (c) Vitreous fluid

    (d) Semen

4. The settling of the blood in the body due to gravity is known as:

    (a) Rigor mortis

    (b) Algor mortis

    (c) Livor mortis

    (d) Vitreous humor

5. Entomology is the study of:
   (a) Fingerprints
   (b) Hair
   (c) Trace evidence
   (d) Insects

6. The mode of death is also known as:
   (a) The manner of death
   (b) The method of death
   (c) The cause of death
   (d) The side of death

7. The decrease of body temperature over time after death is known as:
   (a) Rigor mortis
   (b) Algor mortis
   (c) Livor mortis
   (d) Vitreous humor

8. The cutting made into the torso of a body during an autopsy is known as a:
   (a) X incision
   (b) Y incision
   (c) Z incision
   (d) Prosector

9. A shallow, wide wound with smooth edges might be best classified as a:
   (a) Tear
   (b) Abrasion
   (c) Laceration
   (d) Puncture

10. The sex of a skeleton can be most easily determined with which bone?
    (a) Phalange
    (b) Humerous
    (c) Rib
    (d) Pelvis

CHAPTER 6

Trace Evidence

In forensic science, trace evidence examination is one of the broadest foundations of the applied forensic fields. *Trace evidence* is best defined as the minute or microscopic evidence associated with a crime. When most people think of forensic science, they actually picture trace evidence analysts. The common perception people have of a forensic scientist is that of an analyst in a lab coat, carefully sorting through hairs and fibers, dusting for fingerprints, and examining microscopic debris from a crime scene. In a sense, this perception is accurate, because trace evidence analysis is a major part of forensic investigations. In another vein, though, trace evidence analysis is anything but basic or common. Trace evidence examination is a major area of specialization in forensics, and forensic scientists that choose to specialize in trace evidence analysis go through years of training, practice, and personal certification before they are truly deemed "experts." The reason for this is that interpretation of the results of trace evidence examination is subjective, relying heavily on the experience of the analyst.

In forensic science, the application of the scientific method ensures that all possible factors and outcomes of the tests are accounted for. This is done so that in the end, a hypothesis that cannot be disproved and that is verified by the feedback loop is deemed true within a reasonable degree of scientific certainty. When this process is applied to the identification and individualization of unknown evidence, in most cases there is a clear-cut, unambiguous answer that is backed up by statistical relevance. For example, in DNA testing, the individualization of an unknown DNA profile to a known individual can be assigned a statistical probability based on population genetics. In toxicology, the percentage of a drug in a mixture can be measured, and an estimated error rate can be calculated for that experiment as well. However, certain types of trace evidence, such as hair and soil, are not amenable to statistical analysis or error rate calculations. Analysis of such trace evidence relies solely on the knowledge and experience of the analyst to make a final determination of the inclusion or exclusion of the unknowns to the exemplars in a case.

Even though the examination and testing of the trace evidence is done using the scientific method, the end determination can still be highly subjective. As a result, the expertise of the trace analyst in making the final interpretation must be firmly rooted in the skills, knowledge, and experience that can only be gained after several years of performing trace evidence examinations.

The actual procedures of trace evidence analysis are mainly performed in the laboratory. Although trace analysts can be deployed to a crime scene to aid in the identification and collection of hairs, fibers, and debris evidence, the examination of these items must be done under optimal conditions. Because trace evidence by its very nature is so small, a trace evidence lab needs to have measures in place to ensure that this evidence is not lost or blown away. Therefore, a good trace evidence lab has clean rooms with low airflow, well-lit work areas, and requirements for the analysts to wear proper protective clothing to guard against cross-contamination. In older labs where the environmental controls are not as good, the use of *dead-air hoods* (enclosures that block airflow from disturbing the trace evidence) can be used by the analysts.

The process of trace examination takes skill, time, and patience. It is not uncommon for trace analysts to spend hours picking through batches of hair and fibers, or sorting through dust and soil samples to find representative samples for analysis. Along the way, all of the observations, actions, and results of the tests must be documented. Also, just as in any forensic lab, the evidence itself must be preserved, accounted for, and tracked using a chain of custody. In a trace evidence lab, in which the evidence can be as small as a sliver of broken glass, this can prove to be difficult, and again the skills and know-how of the analyst must be highly developed to ensure the preservation of all types of trace evidence.

Collection Methods of the
Trace Evidence Analyst

Regardless of where trace evidence is found, be it on evidence in the lab, at the crime scene, or on a body at autopsy, it must be collected properly so as not to damage or destroy the evidence, which by its very nature may be small, thin, or fragile. Therefore, a variety of methods can be employed by a trace evidence analyst to ensure the collection of trace evidence is done properly and maintains the integrity of the evidence. These collection methods vary, and can be done by hand using forceps, the adhesive side of a piece of tape, or even a vacuum.

FORCEPS COLLECTION

Due to the miniscule and microscopic nature of trace evidence, some of the most widely used tools in a trace evidence lab are *forceps*, commonly known as tweezers. Depending on the type of trace evidence being examined, there exists a variety of forceps models from which to choose; some lock, some have very light resistance, some have ultra-fine tips, and others are rubberized. Each type has a specific function in trace analysis, but there is no one definite type of forceps for a particular type of evidence. It usually comes down to the personal preference of the trace analyst as to which type of forceps to use.

Trace evidence can be analyzed only if it is successfully located and preserved, and there are several ways of collecting trace evidence. The main process is done with forceps, picking out individual pieces of trace evidence. This can be done at the crime scene, or back in the lab on larger pieces of evidence. For example, if a person's hair gets snagged in the bark of a tree at an outdoor crime scene, it can be collected at the scene. However, if a piece of clothing is collected at the scene, it will be sent to the lab for trace evidence to be collected there. By using forceps to collect individual pieces of trace evidence, the trace analyst can be selective in what they believe to be the most probative elements in the investigation, as well as immediately being able to separate the trace evidence into distinct, single items. However, the downside to this approach is that it is time consuming and requires patience.

TAPE-LIFT COLLECTION

The second method of trace evidence collection uses a broader approach to gathering the evidence with the use of *tape lifts*. The adhesive side of tape is an excellent tool for the collection of fine particles, hair, dust, and other small pieces of evidence (see Figure 6-1).

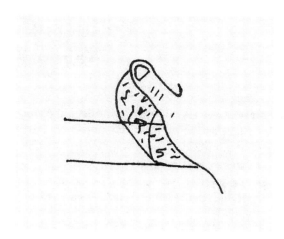

Figure 6-1 Trace evidence on tape lifts

The use of tape lifts is also beneficial in that large surfaces can be quickly covered in a series of lifts, which in theory should collect all of the trace evidence in that area. Tape lifts are commonly used to collect trace evidence from the upholstery and floorboards of a car, and from clothing or other personal items that come into a trace evidence lab. Tape lifts can even be used to collect dust and glass from surfaces. Basically, anything that sticks to tape will be collected in the tape lift.

The tape itself then is submitted to the trace evidence lab, at which point a trace analyst uses a microscope or magnifying glass to examine the adhesive surface and remove any trace evidence that would be probative in the investigation. Unfortunately, tape lifts can pick up a lot of unnecessary debris, hairs, fibers, soil, and whatever else is on the surface that is being taped, and thus is not a selective process. It still comes down to a trace analyst with a pair of forceps to sort through the tape lifts and identify the key evidence that needs to be analyzed.

VACUUM COLLECTION

A third method of trace evidence collection is *vacuum collection*. Vacuum collection is performed using a vacuum source and a special filter that traps all of the evidence that is sucked up by the vacuum (see Figure 6-2). This filter can be a special bag, a filter chamber, or a special piece of filter paper that retains the trace evidence. Vacuum collection is an ideal method for covering a large area quickly to collect trace evidence. However, this method is extremely indiscriminate, and will collect absolutely everything that is present at the scene. Therefore, as with tape lifts, the trace analyst must examine the contents of the vacuum filter to isolate and identify the most probative pieces of trace evidence.

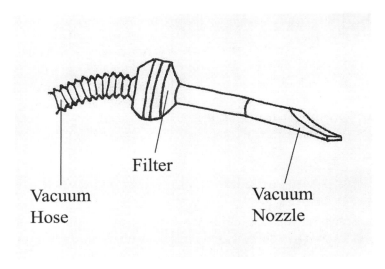

Filter

Vacuum
Hose

Vacuum
Nozzle

Figure 6-2   Diagram of a vacuum filter

The Microscope

Following the collection and separation of trace evidence, a trace evidence analyst
begins the detailed examination of the individual pieces of trace evidence. This task
is usually done with a variety of microscopes. A microscope is one of the most
basic and yet most valuable tools that a trace evidence analyst has at their disposal.
The following sections describe the various components of a microscope and the
types of microscopes that are available:

LIGHT SOURCE (ILLUMINATOR)

Most modern microscopes use a built-in lamp as a light source. This light helps
illuminate the object that is being examined either by transmitting light through the
object (like a thin section on a microscope slide) or by reflecting light off of the
object (as in stereomicroscopy). The bulb of the lamp usually is a broad-spectrum
light and should not be covered with a diffuser such as a frosted lens or "softening"
filter." At higher powers of magnification, the illumination must be uniform, undif-
fused, and at the proper *color temperature*. As mentioned in Chapter 4, color temperature is a measure of the degree of "warmness" or "coolness" of the light source,
and a nominal color temperature is important for high-detail magnification. This
critical illumination, known as *Köhler illumination* (because it was first described
by Dr. August Köhler, in the late 1800s), when properly set ensures that the object
that is being microscopically examined is properly illuminated.

CONDENSER

The condenser is a device that is made up of a series of lenses. In a microscope, the condenser collects and focuses light from the illuminator onto the sample. The proper use of the condenser is necessary in focusing light on a sample for Köhler illumination because it can control the focus of the light in a predictable manner. In most microscopes, the condenser sits below the stage and focuses light from below onto the back of a glass slide. In more advanced microscopes, the condenser can actually be above the stage and transmit focused light down on the sample.

STAGE

The stage of a microscope is a level surface that can hold the sample, whether it is a microscope slide or a larger, three-dimensional object. The stage usually is capable of moving up and down to aid in focusing the sample, and can also move in the other two dimensions to allow a sample slide or object to be scanned by the microscope user.

OBJECTIVES

An objective on a microscope is the device that magnifies the object by using a series of lenses. Most microscopes have a series of different objectives mounted on a rotating ring. Moving this ring changes the magnification of the object that is being observed. The magnification power of an objective lens is measured by the number of times the object is magnified; hence, an objective that is 10x magnifies a sample 10 times. Objective powers can be as low as 0.5x and as high as 100x.

Since objectives are made up of a series of glass lenses, they are susceptible to certain anomalies known as *aberrations*. In terms of optics, aberrations are side effects caused by the bending and focusing of light.

- **Spherical aberration**   Result of the center of the object being magnified to a lower or higher degree than the edges. This is commonly seen in wide-angle or "fish-eye" lenses. If the center of the object is magnified more than the edges, this is known as *barrel distortion*. If the center of the object is magnified less than the edges, this is known as *pincushion distortion* (see Figure 6-3).
- **Chromatic aberration**   Result of light being split into a rainbow of colors after passing through the lens, just like in a prism (see Figure 6-4).

To correct for both types of aberrations, objectives are built with special sets of lenses configured to eliminate or minimize these effects. As a result, objectives can be very expensive, and should be handled with great care.

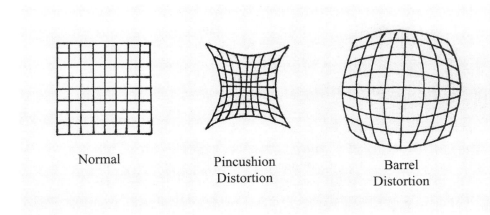

Normal   Pincushion   Barrel
                          Distortion   Distortion

Figure 6-3   Spherical aberrations

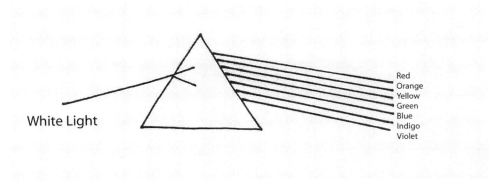

White Light

Red
Orange
Yellow
Green
Blue
Indigo
Violet

Figure 6-4   Chromatic aberration

TUBE LENGTH AND MAGNIFICATION

The tube of a microscope is nothing more than the main body of the microscope that the light passes through to achieve a final, focused image. To achieve a nominal magnification of the image, most microscopes are built with a standard tube length. The magnification of the intended object is dependent on both the distance from the magnified image to the objective, and the distance from the objective to the object that is being magnified. But, because the tube is a set length, the distance from the magnified image to the objective is set as well, and cannot be changed by the user.

Therefore, changes in magnification must be made by changing the distance of the object to the objective. This can be done by changing objectives from a lower power to a higher power. The distance between the object and the objective decreases and the magnification increases!

OCULARS

The ocular, also known as the eyepiece, is the final point of magnification on a microscope. The ocular is what the person using the microscope looks through in order to see the actual magnified image. Older microscopes used only one ocular, but most modern microscopes have changed to a binocular setup. The benefit of a binocular setup is that the depth of an image can be visualized easier with two oculars. The oculars themselves are lenses as well, and usually have a basic power of magnification of 10x. The power of the ocular is very important, because the total magnification of the object is based on the product of the objective magnification power and the ocular magnification power. Therefore, if the objective has a power of 40x and the ocular has a power of 10x, the overall magnification of the object will be 400x (40 multiplied by 10).

Most modern microscopes also have a way of mounting a camera (either film or digital) to the microscope to document what is being examined. This is especially crucial in forensic microscopy, as the images that are collected are ultimately used in the interpretation of the evidence and presented in court.

TYPES OF MICROSCOPES

Microscopes come in different forms and have different applications; therefore, the choice of which microscope to use depends on the job that needs to be done.

Stereomicroscope

The most basic form of a microscope is a stereomicroscope, which uses low-power magnification (usually 10x to 50x) to give a magnified overview of the evidence. Stereomicroscopes actually utilize two separate microscopes that are arranged side by side, with each microscope focusing on the specimen at slightly different angles. This allows each eye to view the image through the two offset microscopes, so that the image formed appears three dimensional. Stereomicroscopes come in handy when sorting trace evidence, especially when removing it from tape lifts or sorting through the vacuum filters. Most stereomicroscopes have some sort of basic illumination, but there is usually no condenser that focuses light onto the sample, because the relatively low magnification does not require high levels of detail. Also, some stereomicroscopes do not have a stage, but rather are mounted on a movable pole called a "boom." This setup allows the stereomicroscope to be moved over a large area and adapt to different locations and configurations.

Polarized Light Microscope

In a trace evidence lab, most microscopy is performed using a specialized microscope known as a *polarized light microscope (PLM)*. A PLM has all the components of a basic microscope, but also incorporates special features that make it unique. One major difference is the presence of an adjustable, high-quality light source. Since Köhler illumination must provide uniform light at the proper color temperature, the illuminator on a PLM can be adjusted to the optimal conditions for proper illumination.

Another major difference is that the condenser on a PLM contains a special filter known as a *polarizer*. A polarizer is a special light filter in which the molecules of the filter are arranged in such a way as to create *plane-polarized light* (see the sidebar, "Polarized Light"). Plane-polarized light has unique properties that can be used in forensic microscopy to uncover hidden aspects of trace evidence, and is especially useful in identifying crystals and fibers. In order to be able to resolve the polarized light, though, an *analyzer* is necessary, and this is usually built into the body of the PLM, somewhere between the objectives and the oculars.

Most importantly, a PLM contains a rotating stage, on which the sample rests, which allows the sample to be rotated to study the interaction of the sample with the polarized light at all angles. A typical PLM also has other components such as *Bertrand lenses*, *compensator slots*, and *swing-in lenses* for the further analysis of images using plane-polarized light. While an in-depth discussion about these components is outside the scope of this book, you can find further information about polarized light microscopes online or in advanced textbooks, such as *Polarized Light Microscopy*, by Walter C. McCrone.

Comparison Microscope

Another type of useful microscope for the trace evidence lab is the comparison microscope. This microscope is also used in firearm and toolmark analysis, and is discussed in detail in Chapter 8. Basically, the comparison microscope uses two separate microscopes but connects the images from both microscopes into one set of oculars. These oculars have the ability to switch from one microscope or the other, or to see a compound image of both microscopes in the same view, where the right side of the image is from the right microscope and the left side of the image is from the left microscope. This setup is extremely useful when analyzing a known and unknown sample simultaneously, and allows individualizing features to be matched up visually without actually contaminating one sample with the other.

Polarized Light

Light, when transmitted from any source, is made up of a series of waves that vibrate in every possible direction. When a ray of light hits a polarizing filter, all of the waves are blocked from passing through except for those waves exactly aligned with the "molecular slits" in the filter. As these light rays pass through in one direction, they remain in that same direction:

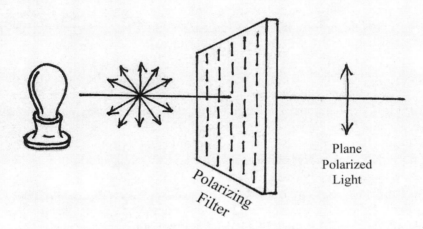

Polarizing filters are present in many forms these days, from window tints to sunglasses. In fact, if you are wearing a pair of polarized sunglasses, you might notice that you cannot read liquid crystal display (LCD) clocks, or see the LCD screens at ATMs. This is because LCDs create polarized light in the process of passing light through the liquid crystals. This light is polarized in one direction, which just happens to be opposite to the direction of the polarizing filters in your sunglasses. However, if you bend your head 90 degrees to the left or right, you may notice that the LCD screen becomes brighter or darker. This is due to the fact that, by rotating your glasses, you are aligning the polarizing filters in the same direction as the plane-polarized light from the LCD screen. When two polarizing filters are arranged at 90 degrees from one another, this is known as *crossed polars*, and no light will be transmitted through the filters. As one of the filters is rotated, light will begin to pass through the filters and become visible, ultimately to the maximum point when both polarizing filters are oriented in the same direction.

ALTERNATIVE TYPES OF MICROSCOPY

There are some other microscopic techniques that can be used by a trace analyst to further investigate the nature of trace evidence. Each of these techniques takes advantage of the optical properties of light and uses light indirectly to study the specimen.

Darkfield Microscopy

Darkfield microscopy is another form of microscopy that is useful in trace evidence analysis. Whereas normal microscopy relies on focusing light onto the sample, darkfield microscopy uses oblique light to illuminate the sample, which means the light is not focused directly on the sample and the objective. Instead, any object in the path of the oblique light "scatters" light toward the objective. The resulting image that is seen in the ocular is bright objects against a dark background. This technique is very useful when looking for dust and other microscopic debris in a sample.

Phase Contrast Microscopy

Another useful microscopic technique in the trace evidence lab is *phase contrast microscopy*. When light travels in a wave form, it has a specific *wavelength* and *amplitude*. The wavelength is the distance from one wave of light to the next, and amplitude is the height of each wave. When all of the light from a source has the same wavelength and amplitude, it is said to be *in phase*. To use phase contrast microscopy, a special plate is inserted into the condenser to create a cone of light that is slightly out of phase with the rest of the light. As this out-of-phase light interacts with samples, any minor differences in optical properties between the samples become even more pronounced. Major differences are then seen as contrast differences between the samples, because the already-out-of-phase light moves even further out of phase by interacting with these samples. This technique requires not only the special plate in the condenser, but special phase-capable objectives as well, and the samples that are being magnified must be small and thin for optimal contrast to occur.

Instrumental Techniques

Alongside microscopic analysis, there are several instrumental tests that can aid a trace evidence analyst. While the individual instruments might not be able to exactly identify what the evidence is per se, they can provide information as to the chemical compounds that make up the evidence, give clues to the structure of the chemicals in the substance, and even detect the elements that are present within the trace evidence. A more in-depth discussion on each of these instruments is provided in Chapter 9, "Toxicology." This section provides a basic breakdown of the instruments that can be used for trace evidence analysis.

ULTRAVIOLET-VISIBLE SPECTROPHOTOMETER

This instrument measures the absorbance of ultraviolet and visible light for a sample. Different chemical structures interact in unique ways with ultraviolet light, and this instrument is capable of measuring the interaction. The ultraviolet-visible spectrophotometer (UV-VIS) is also capable of measuring the color of a solution or extract, and is useful when trying to determine the exact color of a fiber, fabric, or plastic.

INFRARED SPECTROPHOTOMETER

This instrument uses infrared radiation to determine the structure of chemical compounds. Specific wavelengths of infrared radiation are absorbed by certain types of chemical bonds in different ways, and this interaction is measured by the infrared spectrophotometer, otherwise known as an IR. Most modern IR instruments use two forms of data transformation for a more rapid scan of a sample. The first transformation is a physical transformation of the wavelength of the IR radiation by the *Michelson interferometer*. This special device allows all wavelengths of light to quickly scan the sample, which results in a greater energy throughput that can be applied to small samples. Because the scan happens so quickly, several scans can be multiplexed together and the results can be averaged, resulting in the cancellation of noise and background, which ultimately results in higher sensitivity. The second transformation in most modern IR instruments is a mathematical conversion known as the *Fourier transform*. An IR instrument that uses this conversion is therefore referred to as a Fourier transform infrared (FTIR) spectrophotometer.

GAS CHROMATOGRAPH-MASS SPECTROMETER

This instrument is actually made up of two separate instruments that work together. The gas chromatograph (GC) is an instrument that is capable of separating chemical compounds or chemical groups in a mixture. The mass spectrometer (MS or mass-spec) is an extremely sensitive instrument that is capable of taking the separated components from the GC run and analyzing the *mass to charge ratio* of each fragmented component. Based on the GC data and the mass-spec data, a database search can usually reveal exactly what the unknown chemical is that is being analyzed.

OTHER TRACE EVIDENCE ANALYSIS TOOLS

Aside from collection tools, microscopes, and other instruments, some of the most valuable tools in a trace evidence lab are not mechanical instruments at all, but books and online libraries! Because trace evidence analysis looks at small portions of larger objects, there is a good chance that these objects are man-made. When a company produces a new type of fiber, glass, or other material that would be commonly used in everyday goods, information about this material is collected and published in books or gathered in online reference libraries. There are several reference atlases for synthetic and natural fibers, guides to the physical properties of manufactured glass, and indices that outline the chemical properties of thousands of chemicals, powders, and dusts. Online, there are immense computer databases that contain known UV, FTIR, and GC-MS data on all types of chemicals, fibers, glass, and thousands of other natural and man-made objects.

An established trace evidence lab will also have *reference sets* on hand for comparative analysis. These reference sets can be made up of known hairs or hair types, different hairs from animals, local soil samples, glass and fiber standards, and even samples of known common chemicals. All of these tools together can be used by the trace evidence analyst in answering the questions of individualization and identification in the course of a forensic investigation.

Types of Trace Evidence

Once you understand the basic tools and techniques of trace evidence analysis, it is important to take a close look at the types of evidence that can be encountered in a trace evidence lab. Although there are many types and forms of evidence that may be encountered in a trace evidence lab, the methods and tools previously described are universally applied to all types of evidence. Only by careful analysis, observation, interpretation, and application of knowledge can a trace evidence analyst come to a scientific conclusion about the nature and uniqueness of the trace evidence in a case.

HAIR

Hair is one of the most common forms of trace evidence encountered at a crime scene. The average person loses up to 100 head hairs a day. These hairs are simply shed by the body and can leave an indicative trail that can account for the presence of a person at a specific location. Such hairs that fall out naturally are known as *telogen* hairs, and represent the final growth stage of hair. As hairs are actively growing, they are known as *anagen* hairs. At some point, the hair ceases growth but remains embedded in the follicle, and this is known as the *catagen* phase of hair growth. Each of these growth phases has unique microscopic features as well. If an anagen hair was forcibly ripped out of the scalp, there would be an adherence of skin cells known as the *follicular tag* at the root end of the hair, due to its active growing cycle. The catagen phase marks the point where this root begins shrinking and separating from the follicle. The telogen hairs, when found at a scene or on clothing, do not show any sign of the follicular tag, having fully separated from the follicle and fallen out.

These growth differences are occurring randomly to all of a person's hair, all of the time. While these growth phases can cause differences in the nature of the hair on a single person, the physical characteristics of hairs can vary greatly as well on a single person. This can best be seen on a person who has gray, thick hair on the temples but thin, dark hair on the top of the head. If just one of the gray hairs is recovered and compared against the thin dark hair, there of course would be little in the way to match these two hairs to the same person. For this reason, a good *reference set* of hairs is usually required from a person in the course of hair analysis, whether it is the suspect or the victim. To ensure that the reference set is representative of all of a person's hair, several different regions of the head might be plucked and combed.

Hair doesn't grow only on the head, of course. Body hair, also known as *somatic* hair, can be analyzed in the same way. Most commonly, pubic hairs are analyzed by a trace evidence analyst in sexual assault cases. The major differences between somatic hairs and head hairs are thickness of the layers of the hairs, and the structures of the hair. All hair is composed of layers, similar to a tree. The following are the three layers of hair (see Figure 6-5):

- **Medulla**   The innermost layer, it is actually a semihollow canal running the entire length of the hair.
- **Cortex**   The central layer, it comprises the bulk of the hair. Hair gets its color from pigment granules in the cortex.

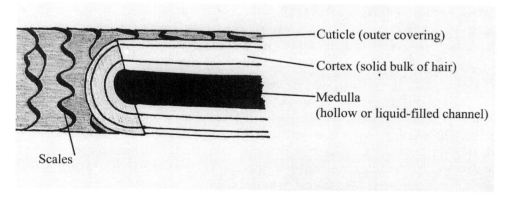

Figure 6-5   The layers of hair

- **Cuticle**   The outer layer, it is actually row after row of microscopic scales, much like the shingles on a roof, except the loose ends point up toward the tip of the hair.

These three layers form as hair grows up out of the follicle in the skin. Each of these layers is formed from hair cells that have hardened and formed into one of the three layers.

During trace examination, a hair needs to be mounted on a microscope slide for microscopic analysis. Hair examination can reveal several key aspects about the nature of the donor of the hair, most notably the species. Since domestic animals such as dogs and cats are so common, it is not unusual to come across animal hair in trace evidence analysis. The scale pattern on the cuticle of a hair is a key factor in determining the species of the donor. Apart from the scales and cuticle, the microscopic traits of hair such as the size of the medulla and cortex, the presence or absence of pigment granules, and other cellular aspects are all examined by the analyst. If these same traits can be found in the reference set of hairs, then a possible inclusion can be established.

FIBERS

Fibers, in forensic terms, can be thought of as the particles and pieces that are used to make fabric, rope, paper, insulation, and hundreds of other materials. Fibers are so prevalent in our environment that if one were to stop and count the fiber sources in just one room of an average house, there might be several hundred to over a thousand possible sources identified. Of course, not all of these materials readily shed their fibers; paper, which is made of wood pulp and (depending on the quality) some form of cotton, does not readily shed its fibers. Nylon and most other woven plastic materials also do not shed fibers. However, an angora sweater or a wool blanket will easily give up loose fibers. This is not to say, however, that the non-shedding fibers will not be discovered as well, and a good trace analyst should be familiar with all types of fibers and their physical characteristics.

As far as classification is concerned, fibers fall into one of three categories (see Figure 6-6):

- **Plant fibers**   Some of the most common and widely used fibers. Some familiar ones are jute, hemp, flax, and, of course, cotton. Each of these materials is made by tightly spinning the fibrous ligaments of specific plants into threads. These threads can then be woven into sheets, or used as threads for sewing. When jute is spun and woven together, the material is known as burlap, and is used extensively in shipping food materials. When flax is spun and woven, it becomes linen. Cotton comes from the fibers surrounding the cotton seeds, and is used in nearly everything.

- **Animal fibers**   Fibers that originate from animal sources. These include silk, from silkworm eggs, sinew, and various forms of hair fabrics. Wool is probably the most common form of animal fiber, with normal wool coming from the hair of sheep. *Alpaca* is another type of wool that is woven from llama fur. *Angora* is a type of wool from regular goat fur, and *cashmere* is a type of goat wool from a specific breed of goats.

- **Synthetic fibers**   The most prolific types of fibers used today in the manufacture of material goods. Synthetic fibers are typically produced by a chemical reaction between two liquids that results in a solid. As this solid is extracted from the reaction mix, it can be formed into any shape or size by pulling it through a *spinneret*. A spinneret contains holes of a specific shape and diameter that end up forming the diameter and cross-section shape of the fiber, similar to the way in which pasta is made. Some of the most common synthetic fibers are nylon, polyester, polyolefin, acetate, and acrylic. Whereas these are generic terms based on the chemical structure of the fibers, several different manufactures over the years have associated trade names to specific synthetic fibers as well, such as Dacron, Kevlar, Spandex, and Nomex. Sometimes, the cross-section of a synthetic fiber is unique to the fiber as well. Most synthetic fibers that are used in outer-layer or single-layer clothing are solid in their cross section. These solid cross sections might have different shapes, from round to square to tri-lobed, depending on the size and shape of the spinneret that was used to form the fiber. Likewise, many synthetic fibers are used for insulation and stuffing purposes, such as the fill used in winter coats or sleeping bags. These fibers tend to have hollow channels running throughout the length of the fiber, which trap warm air and give the fiber more bulk. Some of the common insulating fibers are marketed under trade names such as Quallofil and Hollowfil. Such insulating fibers have unique cross-sections based on the formation of their hollow channels and, when examined, can help in identifying the fiber.

      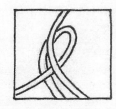

Plant Fibers   Animal Fibers   Synthetic Fibers
                    (hair or fur)

Figure 6-6   Plant, animal, and synthetic fibers

The identification of a fiber in trace evidence is most commonly done using the polarized light microscope. Natural and animal fibers have unique microscopic properties that can be determined both with natural and polarized light. Synthetic fibers, because of the way that they are produced, exhibit unique properties under polarized light. When a synthetic fiber is pulled through the spinneret, the molecules in the fiber are forced to align in the same direction. This creates a pseudo-crystalline arrangement of the molecules, which then interact with polarized light in a particular manner. Under crossed polars, if nothing interferes with the polarized light, no light will pass through the second polarizing filter. However, if a synthetic fiber is placed on a PLM between the polarizer and analyzer, the fiber itself can *rotate* the polarized light from the polarizer and allow some of the light to pass through the analyzer, giving the appearance of a glowing fiber on a dark background (see Figure 6-7). Therefore, under crossed polarizing filters, different synthetic fibers rotate polarized light in different ways, and by measuring the angle of rotation, an identification of the fiber can be made.

Another important identifying characteristic of fibers is whether they are *birefringent*. Birefringence is the "splitting" of a polarized ray of light into two different rays. Materials that are birefringent are usually crystalline in nature, such as mineral crystals and, of course, synthetic fibers. When the material splits the light into two rays, the wavelengths of the two light rays "interfere" with each other, which results in specific *interference colors* that can be seen using a microscope. The birefringence of a material is measured on a PLM based on the interference colors. These measurements, which when correlated against the diameter of the fiber, can be indexed on the *Michel-Levy chart*, a graph that is used to determine the birefringence of a material. The birefringence is a number, which when cross-indexed along with the physical and optical characteristics of the fiber, can be looked up in a fiber atlas for identification.

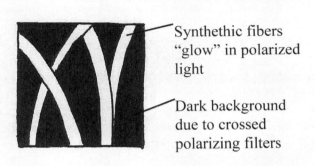

Figure 6-7   An example of synthetic fibers in crossed polarizers

GLASS

Glass is commonly encountered as trace evidence, although it is usually only seen as shards or fragments of broken glass. These small pieces of glass, however, still have the same optical and physical properties as the larger glass item from which they came. Glass is a solid substance, but it is defined as an *amorphous solid*, meaning that it actually is more a slow-flowing liquid than a true solid. Glass also does not have a set crystalline structure. Glass is commonly made up of silicon dioxide, or sand, that has been melted and cooled to form a clear sheet.

The most common type of glass produced is *float glass*, which is made by pouring molten glass onto a molten bed of tin. The glass "floats" on top of the tin and spreads out in a uniform thickness. As the glass moves along the tin bath, it is slowly cooled so that at the end, a solid sheet of glass with a uniform thickness is extruded from the bath, which can then be cut or shaped in any manner necessary. During the float glass process, blowing a supercooled stream of air on the top surface of the molten glass as it is cooling creates *tempered glass*, or safety glass. Tempered glass breaks into cubes rather than sharp shards, because of the stress of having one side of the glass pane cool before the other. Tempered glass is the common glass used in automotive side and rear windows, as well as some building windows and glass tables. Automotive windshield glass is also tempered, but for additional safety it is also *laminated*, or covered in plastic on both sides, which helps contain the windshield glass in case of an accident to keep glass fragments from flying and injuring the driver.

Glass can have other substances added to it to change its physical and optical properties. Most common glass is known as *soda-lime glass* due to the addition of sodium carbonate and calcium oxide to the glass mixture. If lead is added to glass, then a decorative glass known as lead crystal glass is formed. If boron is added to the original glass mixture, then borosilicate glass is produced. Borosilicate glass is also known as Pyrex, and is used extensively in kitchenware and lab glassware. Other minerals and elements can also be added to glass to give color to the glass. Colored glasses are commonly encountered in bottles, but can also appear in glass art and stained-glass windows.

The physical properties of a glass fragment can tell a trace evidence analyst much about the type of glass that was broken. If the fragment is cubed, then there is a good chance that the glass was tempered. If the fragment is colored, the color can be matched to a specific type of glass, whether it is a common bottle color or some other source. The most common characteristics about glass that are commonly investigated in a trace evidence lab, however, are the density of the glass and the refractive index.

The density of the glass can be found by matching the glass chip's density to the density of a similar liquid. If the glass chip is placed in a liquid that is less dense than the chip, the chip will sink. If the liquid is denser, the chip will float. Common glass has a relatively narrow range of possible densities, and if the glass chip is placed into a mixture of chemicals that encompasses that range, the chip will eventually "suspend" in the liquid if the densities are the same. This liquid mixture can then be measured on an instrument to find the exact density of the liquid, and therefore the glass chip. This density can then be referenced against known densities of glass in an index or atlas, and a possible match can be found, which can indicate the source of the glass.

Refractive index, another important optical property of glass, is a measure of the degree that light is bent as it passes through the glass. Different glasses have different refractive indices, and if the refractive index of a piece of glass can be found, it can be matched up with a known source, or identified as a common form of glass. This same phenomenon is seen in swimming pools. If a person standing at the edge of a pool sees a coin on the bottom of the pool, they are really only seeing an image of the coin, and not actually where the coin is. The water in the pool refracts the light in such a way that what appears to be a straight line to the coin is actually a couple of feet in another direction. The refractive index of glass can be measured by surrounding the unknown glass chip in liquids of differing refractive indices. Once the refractive index of the glass matches the refractive index of the surrounding liquid, the glass chip will in fact "disappear," because optically there is no difference between the boundary of the glass and the surrounding liquid. The refractive index of the liquid can then be measured, matched against the glass chip, and cross-referenced in an atlas or index. Once both the refractive index and the density of the glass are known, the glass chip can be identified as a specific type or make of glass.

SOIL

Soil, commonly known as dirt, is actually a mixture of organic and inorganic compounds, and can vary widely between geographical locations. In trace evidence analysis, the mixture of the components of soil need to be separated and analyzed in order to find an identifying feature that could point to an origin. Soil samples can be separated using a *density gradient*, which is a series of increasingly denser liquid chemicals arranged one on top of the next in a glass column. As a soil sample is added to the top layer, the heaviest components of the soil pass through the less-dense upper layers and eventually settle on a layer that matches that component's density. The manner and pattern in which one soil sample separates can be compared against known soil samples to find the same separation pattern.

Sometimes, the chemical components of the soil can be extracted and analyzed using an instrumental technique. A high level of certain contaminates or elevated levels of specific chemicals might point to a source of the soil near industrial areas or processing plants. The organic and biological components of soil can also be microscopically identified, and might contain seeds, bark, leaves, or other components from indigenous plants. High levels of bacteria might point to an urban setting near a farm or common animal source, or some kind of refuse disposal area like a dump.

PAINT

When paint is encountered in trace evidence analysis, it is usually in small chips, peels, or scrapes as a result of one painted object coming in contact with another. Most painted objects have several layers of paint applied to form a finish. Sometimes, there may be a clear top coat of lacquer or sealant, as in automotive paints. During analysis, the microscopic sections of paint can be determined by visual analysis, as well as color determinations by either UV-VIS instrumental analysis or colorimetric matching of the paint samples to known standards. The paint sample can also be extracted and analyzed on an FTIR instrument or a GC-MS to get a chemical analysis and, ultimately, a "signature" of the paint that was found as evidence. If this paint can be matched back to a source, that source can also be tested to see if the same layers and chemical properties are present. In fact, if a paint chip is actually uncovered at a crime scene and is determined as having come from an automobile, there is a good chance that the chip can be matched directly back to a chip in the paint job of the car, proving the source of the chip. This is known as a physical match, or a "jigsaw match."

Fingerprints

Although fingerprints are being addressed in this chapter on trace evidence, they are not, by definition, trace evidence. Fingerprints are not made up of components or particles of larger pieces. However, the techniques and instruments that are used to analyze fingerprints are similar to the techniques and instruments used in trace evidence analysis. The biggest difference between fingerprint analysis and trace analysis is that fingerprints can provide a true individualization of a person, whereas even on normal trace evidence, such as glass, individualization can be difficult if not impossible. Fingerprinting is such a powerful tool for individualization that it can be used to differentiate between identical twins, which is something not even DNA analysis can provide (for an example of this, see the description of the Will West case in Chapter 1). An entire book could be written on fingerprint collection and analysis, so this section gives only a brief overview on the science, techniques, and technology that are used in forensic fingerprint analysis.

Fingerprints, when found on evidence, are actually the negative images of the indentations of the skin ridges on the finger. These ridges fill with oil from other regions of the skin (the tips of the fingers do not have their own oil glands), and when pressed against a surface, this oil is left behind in the shapes of these original indentations. Even after the oil dries, the residue that is left behind still shows the original formation of the fingerprint ridges. This oil or residue can then be dusted, fumed, or developed in such a way to make the print more visible, and then the print itself can be photographed, lifted, and preserved as evidence.

The ridges that make up a fingerprint grow and form into particular patterns, and these patterns are commonly seen and defined in forensic fingerprint analysis. Whenever a single ridge of skin divides into two ridges, this is known as a *bifurcation*. The pattern of the ridges on the print can be classified into *loops*, *whorls*, and *arches*, based on the way the ridges grow, particularly around the center of the fingerprint. Other identifying marks on a fingerprint might be a *delta*, which is a triangular-shaped region of ridges, or an *island*, which is simply a small, single section of ridge between other longer, continuous ridges. All of these classifications are known as *minutiae*, or small identifying components of the larger whole, and they all aid a fingerprint examiner in classifying a fingerprint (see Figure 6-8).

In the past, a fingerprint was classified based on counting the number of ridges between specific patterns, such as identifying the center of the print and counting the number of ridges out to the first major minutiae. This was a time-consuming process and relied on a high-quality development of a fingerprint from a crime scene or an item of evidence. If an unknown print could be classified, then the print could be manually compared against known prints from a suspect, or compared to known prints on file in a police station. The Henry Classification System

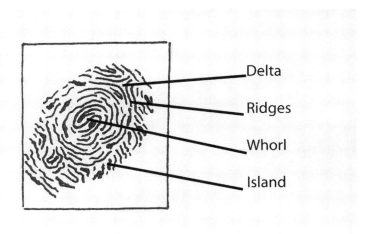

Figure 6-8   Fingerprint showing ridges, whorl, delta, and island

(see Chapter 1) made this cross-referencing an easier task, but the act of comparing the prints manually was still time consuming.

Modern fingerprint analysis is done on computers using image-recognition software and an integrated nationwide system known as IAFIS, or the Integrated Automated Fingerprint Identification System. Individual prints can be quickly scanned into IAFIS using standard computer scanners, photographs, slides, or even print lifts, and these digitized prints are quickly indexed and compared against thousands of similar known and unknown prints. What used to take weeks can now take minutes for an exhaustive search, and the results can then be quickly compared by a trained analyst to make the final confirmation. Typically, 12 to 13 similar minutiae points are needed for a true match to be established. However, recent questions have arisen regarding whether this number of identifying points is adequate for a match, since no empirical studies have been done to confirm this.

Quiz

Refer to the text in this chapter if necessary. Answers are located in the back of the book.

1. What is the microscopic technique in which particles scatter light toward the objective lens?
    (a) Comparison microscopy
    (b) Scanning electron microscopy
    (c) Polarized light microscopy
    (d) Darkfield microscopy

2. Glass is:
    (a) An amorphous solid
    (b) A crystalline liquid
    (c) A cohesive liquid
    (d) A crystalline solid

3. Float glass is made by flowing molten glass over a bed of molten:
    (a) Aluminum
    (b) Magnesium
    (c) Tin
    (d) Mercury

4. Which type of microscopy does not use light?

   (a) Comparison microscopy

   (b) Scanning electron microscopy

   (c) Acute total reflectance microscopy

   (d) Darkfield microscopy

5. Glass can be examined by which of the following methods?

   (a) Fracture matching

   (b) Density

   (c) Refractive index

   (d) All of the above

6. The component of a microscope that collects light and focuses it on the stage plane is the:

   (a) Compensator

   (b) Reticle

   (c) Condenser

   (d) Ocular

7. When a single friction ridge splits into two separate ridges without interruption, this is known as:

   (a) A delta

   (b) A whorl

   (c) Bifurcation

   (d) Both B and C

8. A special type of microscope that allows two separate objects to be viewed simultaneously in the same ocular is known as a:

   (a) Compound microscope

   (b) Comparison microscope

   (c) Dissecting microscope

   (d) Polarized light microscope

9. Which of the following is not a property determinable with a PLM?

   (a) Mass

   (b) Size

   (c) Color

   (d) Birefringence

10. The following components are unique to a polarized light microscope:

    (a) Bertrand lens, stage diaphragm, camera mount, compensator slot

    (b) Rotating stage, polarizer, compensation slot, analyzer

    (c) Optical bridge, 500x objectives

    (d) None of the above

CHAPTER 7

Forensic Biology/DNA

It has long been an axiom of mine that the little things are infinitely the most important.
–Arthur Conan Doyle, *A Case of Identity* (Sherlock Holmes)

Despite the preceding statement, it is doubtful that Sherlock Holmes could have ever predicted that such a tiny little molecule inside our cells would become one of the most powerful tools in the forensic scientist's arsenal. In recent years, however, DNA has become instrumental in the conviction of the guilty, the exoneration of the innocent, and the sorting out of historical mysteries. Beginning with the origins of serology testing in the 19th century through today's automated DNA testing, certainly, it can be said that of all the significant advances made in the field of forensic science, forensic biology has made the greatest and most profound leaps.

Serology

Prior to the integration of DNA testing into the forensic laboratory, scientists who examined evidence for the presence of stains of biological origin were known as forensic serologists. The word s*erology* refers to the science of studying serum, which is the clear yellow fluid found in separated whole blood. Forensic serology, however, also dealt with the analysis of semen, saliva, urine, and other physiological fluids. Therefore, the term *forensic serologist* really fell into disuse, as it did not fully explain the full gamut of work that these forensic scientists did. Today, the field has developed into a much broader one and is known as *forensic biology*. The job of the forensic biologist is to detect, identify, and type body tissues or stains left at a crime scene to ultimately determine to whom they belong.

The detection and identification process starts with a physical examination of the evidence collected at a crime scene. The examination begins with a careful visual examination of the physical evidence and is often coupled with alternate light sources to enhance the detection of any biological stains that might be present. Semen, for example, tends to fluoresce under certain types of light. This is what is being done when a forensic biologist or crime scene investigator uses what looks like a blue light to illuminate a piece of bedding or other item of interest. One important thing to remember is that during this process and all subsequent tests, it is imperative that detailed notes and documentation be kept, as mentioned in previous chapters.

After a stain has been identified, a *presumptive* test, which is a color-change test that screens for the presence of a biological fluid, is performed. These tests are quite sensitive, but not that specific; they do not confirm the presence of a particular fluid. (These presumptive color tests are similar in nature to those used in drug cases, as discussed in Chapter 9, "Drugs and Toxicology.") Bloodstains, for example, often appear similar to rust or ketchup stains. To prevent unnecessary testing, a positive or negative presumptive test can steer the scientist in the right direction in a case to avoid unnecessary consumption of resources later on.

BLOOD

One such presumptive, or screening, test for the presence of blood is the use of *Phenolphthalein*, which is known as the *Kastle-Meyer (KM)* test. Upon the addition of hydrogen peroxide, the presence of hemoglobin (the oxygen-carrying molecule found in blood) causes, or *catalyzes*, an oxidative reaction to occur resulting in a color change from clear to pink.

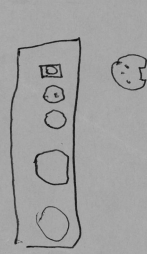

It should be noted, however, that animal blood also results in a positive presumptive result and, therefore, this test does not give a definitive answer for the presence of human blood. Care must also be taken in interpreting presumptive test results for blood, as there is the possibility of false positives and false negatives. The only confirmatory tests for blood are microcrystal tests that produce unique crystals upon the addition of specific chemicals, which are known as the *Takayama* and *Teichman* tests. These confirmatory tests are rarely done in today's forensic labs because once bloodstains are identified, they are usually sent for DNA testing. Since forensic DNA testing is only specific for primate DNA, the likelihood that DNA profiles from nonhuman blood will be used for evidence is miniscule. There is, however, the very remote possibility that gorilla saliva (from a primate) got mixed together with dog blood and produced a DNA profile. One must consider all possibilities in forensic science!

In addition to the KM test, there are several other presumptive tests for the presence of blood that forensic scientists use. *Leucomalachite green* is another example of such a color test; it changes from clear to green when blood is present.

Another presumptive test for blood that is notoriously seen on forensic television shows like *CSI* is the luminol test. *Luminol* is a type of chemical that gives off light in the presence of blood or, more specifically, exhibits *chemiluminescence*. Luminol is generally utilized at crime scenes by spraying it on surfaces such as walls, floors, or other structures where the presence of blood might be present. The visualization of luminol is best seen in the dark, so the lights need to be turned off and the windows need to be covered at a crime scene when performing this technique. Because of its high sensitivity, luminol is often used at crime scenes where it is suspected that the perpetrator has cleaned up. Even in cases where blood has thoroughly been cleaned off of walls, floors, etc., luminol can still pick up small traces of blood that were left behind, such as behind floor boards. Crime scene investigators can then pick up the blood trail, where it might otherwise have gone unnoticed. This can be particularly useful in homicide reconstructions and corroborating suspects' statements. One drawback with luminol, however, is that it gives false positive results with oxidizing chemicals, like bleach. Also, with very dilute stains and the passage of time, even luminol can have its limitations. There are, however, new products, such as BLUE-STAR® FORENSIC latent blood reagent (www.bluestarforensic.com), which have proven to be even more effective in revealing latent blood.

SEMEN

One of the most probative pieces of evidence in a sexual assault case is the presence of semen. It is usually found in or on body orifices, underwear, condoms, or bedding.

Depending on the type of substrate the semen is deposited on, locating it can be easy (as in the case of a dark surface, since it dries as a whitish stain) or more difficult (when it dries on light surfaces). An alternate light source is also often used to locate stains because of semen's fluorescent properties under certain wavelengths of light.

Once a possible semen stain is located, a presumptive test is performed to determine if semen might be present. This is done by looking for the presence of an enzyme called *acid phosphatase (AP)*, which is found in large quantities in semen. A positive presumptive AP test changes from clear to reddish-purple. The results, however, must be interpreted carefully, because the enzyme will lose activity over time and other bodily fluids contain AP, including vaginal secretions.

Once a positive presumptive test for semen is achieved, a confirmatory test is required. Semen is a mixture of fluid consisting of *spermatozoa*, or sperm cells, and *seminal plasma*. The visualization of sperm under the microscope, therefore, is one of the easiest ways to confirm the presence of semen. Another confirmatory test done in the forensic lab is the *prostate-specific antigen (PSA)*, or *p30*, test. The forensic usefulness for PSA was demonstrated by George Sensabaugh of UC Berkeley in the 1970s. This test enables forensic scientists to demonstrate the presence of semen in the absence of sperm, such as in aspermic men or men who have undergone vasectomies. While PSA is found in low levels of other body tissues in both men and women, high levels of the protein are excellent indicators of the presence of semen. The one exception to this is that high levels of PSA can be found in men with prostate cancer.

One final word on rapists who think they can get away with a sexual assault by wearing a condom—they are wrong! Forensic scientists have tests for just about everything. There are now several laboratories that test for condom lubricants using *infrared spectroscopy*. (See Chapter 6, "Trace Evidence" and Chapter 9, "Drugs and Toxicology," for a more detailed description of IR.)

SALIVA

Another type of bodily fluid that is screened for using presumptive testing is saliva. This type of physiological fluid is often present in sexual assault cases where biting or licking has occurred. Saliva screening utilizes a test to detect the presence of *salivary amylase*, an enzyme found in saliva that helps in the breakdown of *saccharides*, or carbohydrates, in the mouth while chewing. Again, care must be taken in interpreting these results because other body fluids can contain low concentrations of amylase, thereby giving false positive results. While saliva itself doesn't contain cells, it does pass through the salivary ducts in the mouth and picks up oral epithelial cells containing DNA.

Rape and Sexual Assault

The psychological trauma of rape or sexual assault that is inflicted on a victim goes far beyond the physical trauma that is caused by the perpetrator. Rape is a heinous crime that is about aggression, control, and humiliating the victim. While no one should ever have to endure such a devastating attack, it is critical that a surviving rape victim immediately undergo a complete sexual assault examination after being attacked. The sooner this exam is done, the greater the likelihood is of apprehending the perpetrator.

The victim should be treated by a doctor or a sexual assault nurse examiner that has experience performing these types of exams in a hospital. During the exam, in addition to being treated for any physical trauma and/or sexually transmitted diseases, a sexual assault kit is collected. Most kits include a medical record form, a body diagram to document injuries, an authorization for release of information and evidence to law enforcement, a patient information form, and about a dozen different envelopes for collecting evidence. These envelopes are included for the collection of oral swabs and smears, trace evidence, underwear, debris collection, dried secretions and/or bite marks, fingernail scrapings, a control sample of pulled head hairs, pubic hair combings, a control sample of pulled pubic hairs, anal swabs and smears, vaginal/penile swabs and smears, and a saliva control swab. These items of evidence, when properly collected, can yield a large degree of information about the assailant. In addition to the actual kit, the victim's clothing, any condoms used in the attack, and any bedding can also be submitted to the crime lab for analysis. Victims' cooperation with law enforcement in providing possible locations of biological evidence can be invaluable. Moreover, it is also critical that law enforcement provide this information from the victim to crime lab scientists.

DNA Typing

Once a biological fluid has been identified, the sample needs to be typed to determine its origin (see Figure 7-1).

Typing is the detection of a person's genetic profile, which in the case of DNA is represented by a string of numbers. The difference in type between the evidence stain and an *exemplar* (a known reference sample from either the victim or the suspect) provides a positive elimination. If there is no elimination, the results must be evaluated against population frequency data to give a measure of the significance of the failure to exclude. In other words, how statistically relevant is it that two DNA profiles are included in a "match"?

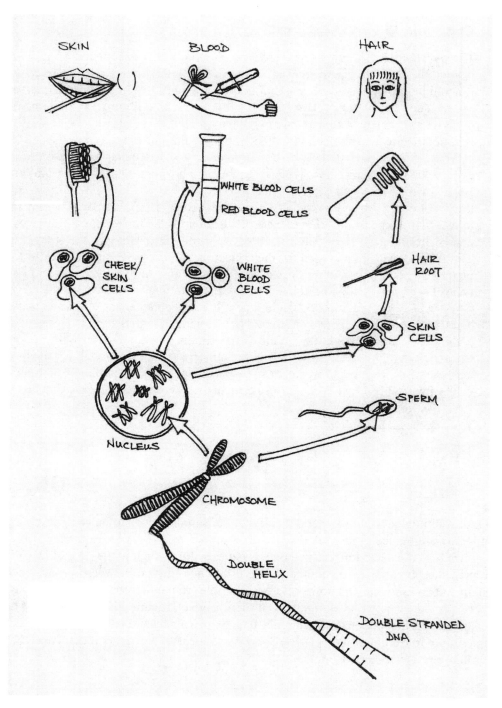

Figure 7-1 Overview of DNA typing

Some of the features that make DNA such a good typing system are the following:

- It shows variability from person to person but is constant within one individual throughout their lifetime.
- It is stable in shed form.
- It can be detected at small concentrations found in forensic samples.
- Its frequency of occurrence within the population is known and stable.

WHAT IS DNA?

DNA stands for *deoxyribonucleic acid*. It is the biological blueprint of life. The DNA molecule in its native form is double stranded and looks like a twisted ladder, or *double helix*. It is a continuous chain of repeating units made up of a sugar (*deoxyribose*) and a phosphate backbone attached to four chemical building blocks called *nucleic acids* or *nucleotides*. The sugar and phosphate group is what provides the handrail for the twisted ladder; the nucleic acids are what provide the rungs for the ladder. These rungs are made up of *purines* and *pyrimidines*. The purines consist of *adenine (A)* and *guanine (G)*. The pyrimidines consist of *cytosine (C)* and *thymine (T)*. Adenine always pairs with thymine, and guanine always pairs with cytosine (see Figure 7-2).

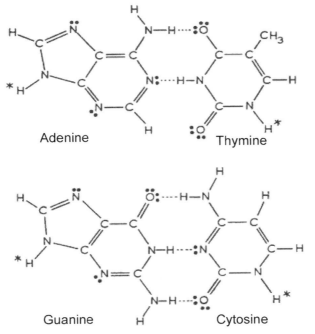

Figure 7-2   The molecular structure of the cross-linking nucleic acids

The human body is comprised of roughly three trillion cells. Inside most of these cells exists the blueprint, or instruction manual, in the form of DNA in the cell's organization center, the *nucleus*. (Red blood cells, which carry the oxygen and make blood red, are so specialized that they do not require a nucleus or DNA.) It is this DNA that is responsible for such things as eye color, straightness of hair, and a host of other physical characteristics.

The DNA inside the cell's nucleus is found in the form of long strands known as *chromosomes*. Human beings have 23 pairs of chromosomes, or 46 in all. Half of your chromosomes come from your mother and half come from your father. Along the chromosome are regions of DNA called *genes*, which act as units of heredity. Alternative forms of the same genes are called *alleles*. ABO blood types, for example, illustrate different alleles in people. People are type A, type B, type O, or, the most rare, type AB. The combination of alleles a person has determines their blood type. The location on the chromosome occupied by the allele is called a *locus* (or plural, *loci*). When two alleles at a locus are the same on both chromosomes, a person is *homozygous* for that trait. If they are different, that person is *heterozygous*. For example, a woman with type B blood can be either homozygous for the B allele (BB) or heterozygous (BO). Both allele combinations, however, still give her type B blood.

DNA EXTRACTION

The first step in DNA typing after a biological stain has been identified is to isolate the DNA from the cells. This can be done in a number of different ways depending on the type of case involved, the amount of sample available, the type of specimen, and whether it is a biological stain, tissue, or bone. (Many people think that forensic DNA analysts work with powerful microscopes to "see" the DNA. This is a common misconception. The majority of work that is done is performed in small tubes with clear liquids where the forensic biologist is unable to "see" what is taking place.)

Because the cell contains many different components, the DNA needs to be separated from the other parts of the cell before it can be examined. This can be achieved in several different ways. Two of the most popular extraction methods are the *Chelex®* method and the *organic* method.

Chelex Extraction

In the Chelex extraction method, heat and water are used to break apart the cell membranes and destroy cell proteins. Chelex, which is a suspension of tiny resin beads, protects the DNA from DNA-destroying enzymes, called *nucleases*, by rendering them inactive. The last step involves spinning the tube of DNA and Chelex in a very fast machine called a *centrifuge*. The centrifuge's g-forces pull the heavier Chelex resin and cell debris to the bottom of the tube, while the separated DNA remains at the top of the tube in what is called the *supernatant*. This type of extraction works very well for forensic samples, such as bloodstains, *buccal* (saliva) swabs, and cigarette butts.

Organic Extraction

Organic extraction uses organic solvents, hence its name. In this method, chemicals are first added to break open the cell wall and to destroy proteins that protect the DNA. Next, phenol/chloroform and isoamyl alcohol are added to separate the proteins of the cells from the DNA. When spun at high speeds in a centrifuge, the unwanted proteins and cell debris are separated from the DNA into two layers. While this extraction method works well in recovering lots of DNA from difficult samples, it requires the use of hazardous chemicals, and is more time consuming than Chelex extraction.

DIFFERENTIAL EXTRACTION

The Chelex and organic extraction methods work quite well when only a single source of DNA is present. What happens, however, when a mixture of DNA from two individuals is submitted, as is often encountered in sexual assaults? How can an accurate DNA profile be generated?

Mixtures present a unique challenge to the forensic scientist. Differential extraction is one technique that is used to separate sperm cells from *epithelial*, or skin cells, as is often necessary in sexual assault cases. Once the male sperm fraction is separated from the female epithelial cell fraction, it is much easier to obtain the DNA profile from the rape suspect. This type of extraction exploits the morphology, or physical characteristics, of the sperm cells of the male to separate them from the epithelial cells, usually from the female (see Figure 7-3). (In the case of sodomy where the victim is male, the epithelial cells would be from a male.)

The differential extraction of a mixture of cells uses different chemicals at different times. The procedure begins by first breaking open the epithelial cells and isolating the female DNA, followed by breaking open the sperm cells and isolating the male DNA.

As shown in Figure 7-3, the first step the forensic scientist performs is to cut the swab, or stained garment, and place the cutting into a tube (the swab is usually obtained from the sexual assault kit taken at the hospital from the rape victim). Chemicals are then added to the cutting and the tube is shaken vigorously to free the sperm cells and epithelial cells from the substrate. The *substrate remains* (i.e. the cotton swab or stain) is then removed from the original tube and placed into a clean

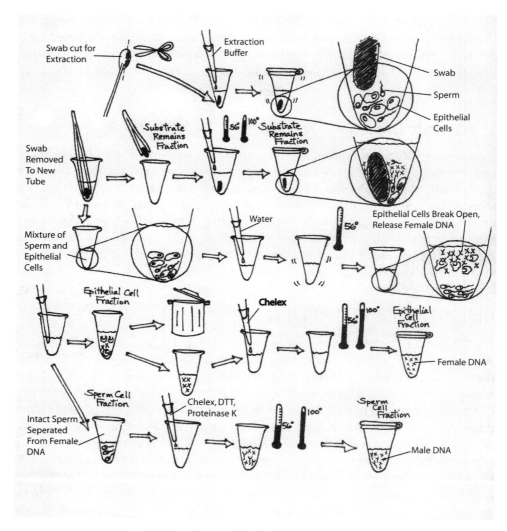

Figure 7-3 Differential extraction overview

tube. This becomes the *substrate remains fraction*. Since not all of the cells will come off of the swab, it's expected that a mixture of female and male DNA will be found in this fraction. This mixed DNA profile from the sperm cells and epithelial cells can be used later if no clean male DNA profile can be obtained.

The next step in the differential extraction (rows 3 and 4 in Figure 7-3) involves adding chemicals to the mixture of sperm and epithelial cells that were removed from the swab or stain, to preferentially break open only the epithelial cells in the mixture. During this step, the sperm cells remain intact with their DNA enclosed, because the nuclear membranes of sperm are tougher than their epithelial counterparts. The female DNA can then be isolated as the *epithelial cell fraction*.

The final step in the differential extraction is to *lyse*, or break open, the remaining sperm cells, thereby freeing the male DNA contained in the sperm. This produces the *sperm cell fraction*.

These three separate fractions—substrate remains, epithelial cell, and sperm cell—then become subsamples of the original sample, and can be used by the forensic biologist to obtain or interpret the DNA profiles of the victim and the perpetrator. While most forensic laboratories use some form of a differential extraction procedure, there are several variations on the protocol depending on the lab.

Occasionally, the components of a mixture cannot be fully separated from one another in a differential extraction, and one is left with a mixture of DNA, as in the case of the substrate remains fraction. Forensic DNA analysts, however, can still deduce the major and minor contributors in a mixture case based on the ratios of DNA present. (A full explanation on the topic of mixture interpretation is beyond the scope of this book.)

DNA QUANTITATION

After the DNA is extracted from the cell, the forensic scientist needs to determine how much DNA is present. This is important for the subsequent steps in the DNA typing process since they require precise amounts of DNA to be used. In quantifying the DNA, the analyst is also able to make certain that the DNA recovered is of primate origin, rather than from another source such as lower-order animals or bacteria. One common method of DNA quantitation in today's forensic laboratories is known as the slot-blot technique, which uses the *QuantiBlot® Human DNA Quantitation Kit* by Applied Biosystems. In this procedure, the unknown sample is compared to known standard amounts of DNA to determine its quantity. Recently, however, many laboratories have been switching to a different quantifying technique called *real-time PCR*, which determines the quantity and quality of DNA present in a sample.

THE POLYMERASE CHAIN REACTION

The next step in the DNA typing process is a procedure called the *polymerase chain reaction (PCR)*. PCR is a method that enables a small amount of DNA to become amplified, or copied, into millions of additional copies. The technique, discovered by Nobel laureate and surfer Kary Mullis, essentially works like a Xerox machine; target DNA can be multiplied into millions of copies of the exact same DNA within a short period of time. The process involves repetitive cycles of heating and cooling of samples in a precise, temperature-controlled environment, which is regulated by a machine called a *thermal cycler*. Each cycle of PCR involves three distinct steps that are repeated over and over again until the desired amount of DNA is reached (see Figure 7-4).

The first step in a PCR cycle is called *denaturation*. During denaturation, the DNA strands separate from one another at a very hot temperature of 94°C. The next step is called *annealing*, which cools the DNA to 60°C. During annealing, small fragments of DNA, called *primers*, bind to the DNA *template* and target the region of interest of the DNA to be amplified. The final step, *extension*, extends the primers on both strands by copying the target region of DNA with DNA building blocks, and occurs at 72°C.

At the end of the first cycle, two copies of the target DNA are generated. When this cycle is repeated, the number of targeted DNA molecules doubles. Usually, an entire PCR run is 25 to 35 cycles and takes 2 to 3 hours. If the average PCR is 28 cycles, then the yield of DNA will be millions of copies (2^{28}).

STRs

In the mid-1980s, Sir Alec Jeffreys discovered that every person's DNA is unique. He and his colleagues found that certain areas of the long human DNA molecule are *polymorphic*, or exist in different forms between one person and the next. These variable areas on the DNA molecule allow forensic scientists to *individualize* people, or distinguish people from one another.

In the human genome, our DNA is full of repeated sequences. These sequences are encountered both in different lengths and different numbers of repeats. If one visualizes a train as being representative of our DNA, different trains can have different numbers of cars, and different sizes of cars. DNA regions with repeat units that are used in human identification are called *short tandem repeats*, or *STRs*. They are *short* because they average four base pairs in length. They are *tandem* because they occur next to each other. And they *repeat* because the same base pair unit is repeated multiple times in the DNA sequence. What is significant about all of this is that the STR markers chosen by the forensic community are highly variable between people and can be used to discriminate you from every other person in the world. This means that, just as the old adage says, you are unique, just like everybody else! The exception, of course, is identical twins, who have the same DNA. (As a result, there have been accused murderers who have pointed the finger at their long-lost twin brother.)

The forensic DNA community has developed a common naming system for these STRs to allow laboratories to share this DNA information with one other. This allows labs across the world to communicate in a common language with regard to

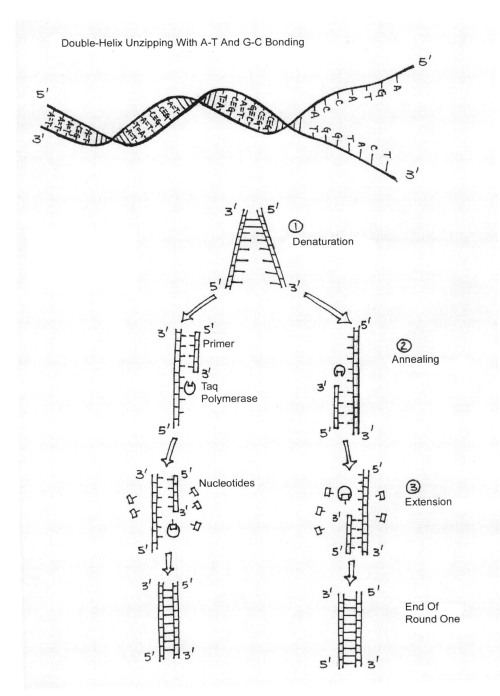

Figure 7-4   PCR amplification overview

DNA. There is currently a standard set of 13 core forensic STR *loci*, or locations, that are used across the United States. Systems in other countries, such as the system used in Great Britain, use some of these 13 loci, in addition to other loci. Today, there are commercial kits that labs can purchase that target and amplify these standard sets of core forensic STR loci. The 13 core forensic loci currently used in the United States are the following: D3S1358, D16S539, THO1, TPOX, CSF1PO, D7S820, vWA, FGA, D8S1179, D21S11, D18S51, D5S818, D13S317. Aside from these 13 locations, STR tests also screen for the gender-determining locus, *amelogenin*.

How does all of this work? After the PCR amplification process previously described, there is a mixture of DNA molecules (different STR markers) that require separation to make sense of all this information. In other words, in any given PCR reaction, there are many DNA fragments that need to be resolved from one another. The separation step that does this is called *electrophoresis*. Because DNA is a negatively charged molecule, it has an affinity for a positively charged electrode. (Remember, opposites attract.) When an electric field is applied to a DNA mixture, the DNA starts to migrate across the field from the negative end toward the positive end. The DNA mixture contained in each sample is added to a gel with a series of small pores that allows the DNA fragments to migrate through it. The smaller DNA fragments tend to migrate through the gel faster than the larger DNA fragments.

This process can be performed in two different types of electrophoresis environments: *slab-gel electrophoresis* and *capillary electrophoresis*. Both types of systems are used in forensic labs, with the trend moving more toward the capillary systems.

In the slab-gel electrophoresis system, wells are made in a solid gel for loading DNA samples. The gel is then placed in a buffer solution. Each sample is then loaded into a different well. Known and unknown samples are never loaded on the same gel, to prevent contamination and sample mix-up. Next, an electric current is applied to the gel and the DNA fragments begin to migrate. A very expensive instrument called a *Genetic Analyzer*, made by Applied Biosystems, then "reads" these separated fragments using a laser. During the PCR step, the DNA fragments are tagged with different-colored fluorescent dyes that allow the laser to differentiate between the different DNA fragments. Computer software finally takes this information and converts it into bands (similar to the supermarket barcodes) and peaks. These peaks are assigned numbers based on the number of repeat units in the DNA molecule. Taken together, these peaks and numbers make up an *electropherogram*, or DNA profile (see Figure 7-5).

In the capillary electrophoresis system, instead of using a slab of gel, a very thin tube called a capillary is used to perform the electrophoresis, or DNA separation. A *polymer*, which is a gel-like compound with tiny passageways, is pumped into the capillary, allowing the mixture of DNA fragments to be separated based on size. A laser then also "reads" these DNA samples and software converts this data into

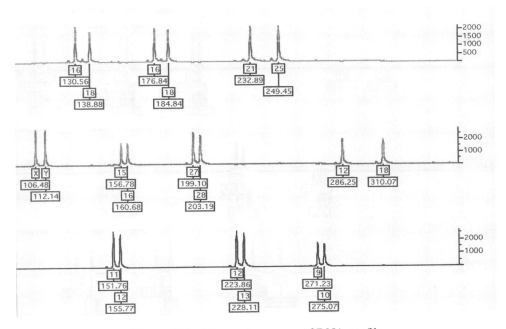

Figure 7-5 Electropherogram of DNA profile

electropherograms. The advantage of the capillary system over the slab-gel system is that the capillary system is faster and more samples can be run at the same time (16 samples can be run in 45 minutes).

Specialized DNA Testing

In some cases additional, or specialized, DNA testing might be warranted. When there are multiple male suspects involved in a case, when an *autosomal* (non-sex determining) DNA profile could not be obtained, or when there is too little DNA present, other DNA tests can aid in the investigation.

Y STRs

Y STRs are found only in males, because males possess an X and a Y chromosome, whereas females have two X chromosomes only. Since the majority of crimes in which DNA evidence is used involve sexual assaults, and since the overwhelming majority of sexual assaults involve male perpetrators, DNA tests that can exclusively examine the male DNA portion are quite beneficial. Cases in which Y STR testing might be used are those that involve very high levels of female DNA mixed with small amounts of male DNA, or where the number of men involved in a "gang rape" is unknown.

While Y STR testing can prove quite valuable in certain types of cases, the results are not as meaningful as regular, or *autosomal*, STRs. This is due to the fact that the Y chromosome is transferred directly from father to son. A match between evidence and a suspect only means that the suspect could have contributed to the sample, as could have his father, brother, son, uncle, or even long-lost paternal cousin. The random match probability, therefore, is much lower for Y STRs and provides a lower power of discrimination between men.

There are also other popular uses for Y chromosome testing, including tracing human migration patterns, researching genealogies, and proving ancient genetic links, such as the link between modern day Cohanim to the ancient Jewish priesthood beginning with Aaron (Exodus 40:15).

MITOCHONDRIAL DNA

Mitochondrial DNA (mtDNA) is another type of DNA that is useful in forensic investigations. This type of DNA is found in another part of the cell, called the *mitochondria*. The mitochondria is responsible for energy production within a cell. MtDNA is less susceptible to degradation than nuclear DNA, due to the fact that it is much smaller, is protected inside the double membrane of the mitochondria, and exists in a circular genome (see Figure 7-6), unlike the linear double-helical structure of nuclear DNA.

MtDNA molecules are also present in thousands of copies per cell, unlike its nuclear DNA counterpart, which has only two copies. Therefore, biological evidence that has been subjected to harsh environmental conditions may yield successful mtDNA typing in cases where nuclear DNA typing has failed.

Another feature of mtDNA is that it can be used to determine ancestry, because it is inherited maternally. This means that you have the same mtDNA type as your mother, who has the same mtDNA type as her mother, and so on. (Nuclear DNA is inherited in equal parts from both parents.) This is because mtDNA is inherited only from the egg cell of the mother. When a sperm cell fertilizes an egg, only the nuclear DNA contained in the sperm head, and not the mtDNA contained in the "neck" of the sperm, makes it into the egg. Thus all the cell components of the developing fertilized egg come from the mother, including the mitochondria. This trait is particularly useful is missing person cases, because a maternal relative can be used as a reference sample in place of the missing person. MtDNA is also extremely stable and undergoes very few mutations, which means that you can accurately trace your maternal lineage back many generations.

Figure 7-6 mtDNA genome

The Romanovs

The Romanovs, the ruling dynasty of Russia from 1613 to 1917, lost power during the Bolshevik Revolution in 1917. Czar Nicholas II, his wife Alexandra, and their five children were sent into exile by order of the Bolshevik leader, Vladimir Lenin. Late one night in July 1918, their imprisonment ended when the entire family was executed by firing squad, along with their doctor and three servants. This act ended 300 years of Romanov rule.

In 1991, during the collapse of the Soviet Union, researchers exhumed nine bodies near Yekaterinburg, Siberia. Nuclear DNA testing performed by Great Britain's Forensic Science Service showed that five of the nine skeletons were from a single family, but which one? Tests then turned to mitochondrial DNA, which is passed down through the maternal line. Great Britain's Prince Philip, husband of Queen Elizabeth and Czarina Alexandra's grandnephew, provided a DNA sample that confirmed that the remains were those of Nicholas II, Alexandra, and three of their children.

Scientists at the U.S. Armed Forces DNA Identification Lab (AFDIL) confirmed that the bodies in the grave were those of the czar, his wife, three of their four daughters, and four other people—presumably the doctor and the three servants. The son, Alexis, and daughter, Anastasia, were rumored to have survived the assassination and escaped. In 1994, DNA testing revealed that the most famous Anastasia impersonator, a woman named Anna Anderson, was in fact not Anastasia. The whereabouts of Anastasia and Alexei remain unknown.

The mitochondrial genome contains about 16,500 bases and contains a *noncoding hypervariable control region*. Within this region, two segments, called *hypervariable region 1 (HV1)* and *hypervariable region 2 (HV2)*, tend to mutate at a higher frequency than the rest of the molecule. It is for this reason that this region is of particular significance for forensic individualization.

Just like Y STRs, however, mtDNA has a low power of discrimination. The probability of mtDNA being unique is usually only around 1 in several thousand depending on the size of the database being used to calculate the number of times that the mtDNA profile appears in the population. MtDNA is, however, an excellent way of excluding a person from being the source of a particular sample.

LOW COPY NUMBER DNA

A new breakthrough in DNA testing that has been introduced into crime labs is the ability to obtain DNA profiles from very small amounts of sample. This type of testing is usually reserved for samples that contain only a handful of cells. Because of the small amount of DNA present in these types of samples, additional PCR cycles are run. To this end, some jurisdictions have begun analyzing property-related crimes like burglaries and robberies where only skin cells might be present. A burglar or robber that handled a stolen piece of property might have left partial or smudged fingerprints containing epithelial, or skin, cells from their hands or fingers. This evidence can then be submitted to the forensic lab for testing. A word of caution, however: low copy number (LCN) DNA results need to be interpreted more carefully since the sensitivity of the DNA typing is so much higher, and thus the chance of contribution from contamination is greater.

Statistics

Statistics are an integral part of forensic DNA. Without reporting statistics, the results of a "match" would be insignificant. For example, if a scientist were to testify to the fact that the suspect has type A blood and the blood found at the crime scene was also type A, what conclusion is the jury supposed to draw from this? Hopefully, not much, because approximately 1 out of 4 people have type A blood.

DNA evidence, however, has much higher statistical weight. The statistical weight of a full 13-locus profile is generally greater than 1 in a trillion. This means that if one were to choose a person at random from society, the odds of that person having the same DNA profile in question would be greater than 1/1,000,000,000,000. Since the population of planet earth is approximately 6 billion (6,000,000,000), it would take around 166 planet earths to obtain a trillion people. To put it a different way, the odds of you winning the lottery are much greater than finding another person with your exact same DNA.

The reason that these odds are so small is that the probabilities of the individual 13 profile types are independent of one another, since the 13 core loci are located on different chromosomes. This means that the individual odds can be multiplied together to produce staggering odds. Statistics for mtDNA and Y STR DNA profiles require that a counting method be used instead of multiplying probabilities because the individual DNA markers are not independently occurring. The counting method produces smaller odds that a profile is unique.

CODIS

The Federal Bureau of Investigation's *Combined DNA Index System (CODIS)* uses forensic science and the power of computers as a tool to solve crimes. CODIS allows local, state, and federal crime labs to exchange DNA profiles electronically in hierarchical levels (see Figure 7-7).

CODIS enables different jurisdictions to link crimes to each other and to convicted offenders. DNA profiles begin at the local level, with local crime labs generating DNA profiles for the *Local DNA Index System (LDIS)*. Local laboratories then upload their DNA profiles into the *State DNA Index System (SDIS)*. Currently, all 50 states participate in the CODIS system. States then upload their profiles into the *National DNA Index System (NDIS)* maintained by the FBI. According to the FBI's website (www.fbi.gov), as of April 2006, the NDIS database has 3,275,710 DNA profiles, including 136,672 forensic profiles and 3,139,038 convicted offender profiles.

CODIS allows scientists and investigators to compare a DNA profile with convicted offenders and other crime scenes. This is important because it can help link cases to one another across jurisdictions that may otherwise have gone unsolved. As the database grows in size, the ability for law enforcement to solve crimes will also increase. As of February 2006, CODIS has produced over 30,000 hits assisting in more than 31,700 investigations. In addition to the NDIS database, CODIS now also includes a missing persons database (CODISmp) and a mitochondrial DNA database (CODISmt).

Other countries throughout the world have also developed, or are developing, similar national databanks that operate like CODIS. Each country, however, has different statutes regarding who is required to submit a DNA profile, what is required for a profile to be expunged from the database, and which STR loci make up a DNA profile. To this end, INTERPOL (International Criminal Police Organization) has been working on the standardization and internationalization of DNA profiling

Forensics Demystified

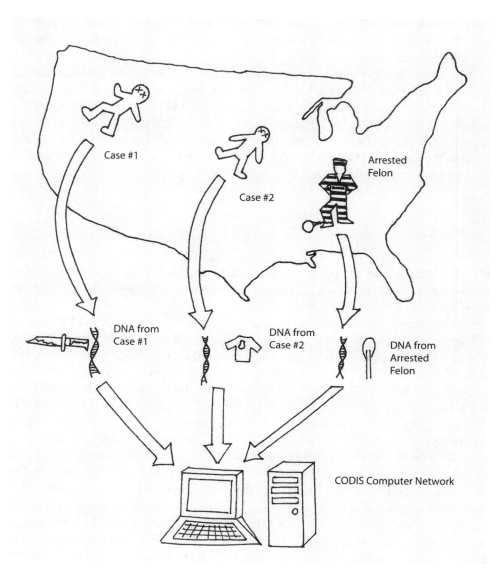

Figure 7-7 Combined DNA Index System (CODIS)

worldwide. As a result, INTERPOL has been able to assist in international investigations involving national DNA databanks from multiple countries.

Mass Fatality Identifications/ Missing Persons

The DNA identification efforts for the victims of September 11, 2001 have become the largest forensic investigation in history. On that day, at the World Trade Center in New York City, 2749 people were murdered, leaving approximately 20,000 pieces of human remains. The goal was and still is to identify as many of those remains as possible in order to return them to their loved ones. Without the capabilities of DNA testing, however, this would largely have been an impossible task and the majority of victims would not have been able to be identified.

Analyzing samples from the World Trade Center site was only half of the process. The other part of the ID effort required the analysis of DNA samples provided by families. This included DNA of the victims from toothbrushes, razors, hairbrushes, and the like. In the absence of these, mouth swabs from biological relatives were used in a process called *kinship analysis* (see Figure 7-8).

By obtaining the DNA profiles of the mother and father of the deceased/missing or the DNA profiles of a spouse along with the victim's son or daughter, identifications were made by applying rules of genetic inheritance and statistics.

Since a more detailed account of the World Trade Center identification efforts are beyond the scope of this chapter, you are encouraged to read *Who They Were: Inside the World Trade Center DNA Story: The Unprecedented Effort to Identify the Missing* (Free Press, 2005), written by the former director of the Forensic Biology Department for the New York City Medical Examiner's Office, Dr. Robert Shaler.

Forensic DNA Testing and Beyond

Forensic DNA evidence can link a suspect to a crime, exculpate falsely accused suspects, recognize serial crimes and distinguish copycat crimes, identify the remains of victims, and reconstruct crime scenes. Parentage testing is a form of identity testing that has increasingly turned to DNA testing in recent years. People researching their genealogy have also turned to DNA typing to help establish a link to their past. History is also being rewritten based on the DNA testing surrounding historic figures, such as Jesse James, the unknown soldier at the Tomb of the Unknown, Czar Nicholas II, Thomas Jefferson, and many others. Forensic DNA testing has come quite a long way since its inception and will continue to make giant strides in the future.

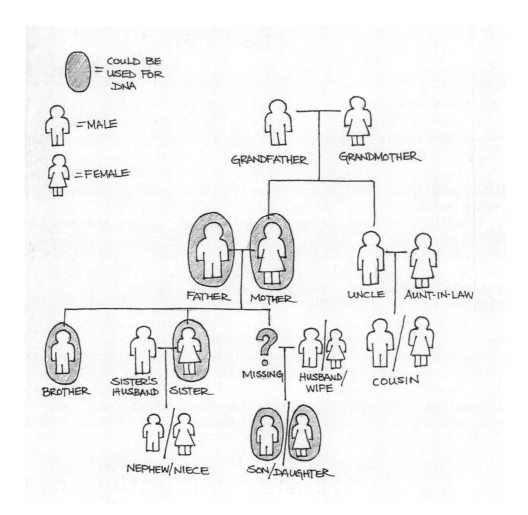

Figure 7-8 Kinship analysis overview

Many of the details in forensic DNA typing are beyond the scope of this chapter. You are encouraged, however, to learn more about DNA from the website www. dna.gov, which was established by the President's DNA Initiative, in conjunction with the National Forensic Science Technology Center.

Quiz

Refer to the text in this chapter if necessary. Answers are located in the back of the book.

1. The source of the p30 enzyme found in semen is:
   (a) The liver
   (b) The testes
   (c) The prostate
   (d) The vas deferens

2. Spermatozoa can be visualized on microscope slides to prove the presence of:
   (a) Blood
   (b) Saliva
   (c) Semen
   (d) Feces

3. Purines are:
   (a) Adenine and thymine
   (b) Thymine and guanine
   (c) Cytosine and guanine
   (d) Guanine an adenine

4. Amylase is found in high concentrations in what stains?
   (a) Semen
   (b) Blood
   (c) Urine
   (d) Saliva

5. The rarest blood type is:
   (a) A
   (b) B
   (c) AB
   (d) O

6. Thymine is a:

  (a) Purine

  (b) Pyrimidine

  (c) Polymer

  (d) Phosphatase

7. The reaction of luminol with blood is best described as a:

  (a) Antigen-antibody reaction

  (b) Cold-fusion reaction

  (c) Sympathetic reaction

  (d) Chemiluminescent reaction

8. CODIS is an acronym for:

  (a) Convicted Offender DNA Index System

  (b) Combined DNA Identification System

  (c) Convicted Offender DNA Identification System

  (d) Combined DNA Index System

9. Pyrimidines are:

  (a) Adenine and thymine

  (b) Thymine and guanine

  (c) Cytosine and guanine

  (d) Thymine and cytosine

10. Which of the following is a positive presumptive test for blood?

  (a) Luminol

  (b) Leucomalachite green

  (c) Phenolphthalein

  (d) All of the above

CHAPTER 8

Firearms and Toolmarks

Crimes involving the use of firearms seem to be ever more prevalent in today's society. Firearms are regularly used in homicides, suicides, robberies, assaults, and other violent crimes. Oftentimes, the perpetrator leaves bullets, cartridge casings, or even the gun itself at the crime scene. These items of evidence, when examined properly, can yield a great deal of information about what took place at the scene of the crime. Several questions can be answered with careful analysis of the remains of shooting evidence. What kind of weapon was used? Was the gun fired at close range? Can this shooting be linked to any others? Moreover, firearms evidence is particularly important in corroborating statements made by witnesses, victims, and/or suspects.

The field of forensic firearms identification is sometimes mistakenly called ballistics. This is really a misnomer. The term *ballistics* usually refers to the science of physics that is applied to the trajectory taken by a bullet or projectile. Forensic *firearms identification* is primarily concerned with determining if a certain bullet or cartridge casing came from a particular gun. Additionally, firearms examiners examine powder patterns, restore obliterated serial numbers from recovered weapons, and often conduct toolmark examinations. *Toolmark analysis* is another subsection of forensic science that deals with the impressions or damage made by tools used in a crime. For example, when a crowbar is used to pry open the door to a safe, the impressions and indentations in the metal of the safe can be matched to the crowbar if it can be recovered from the perpetrator. Toolmark and firearm analyses are usually grouped together because of the general practice of pattern matching that is done in each of the fields.

Firearms

Most guns are usually classified into one of three different categories: *handguns*, *rifles*, or *shotguns*. The differences of each type of firearm are based on the size of the weapon, the type of ammunition that is used, and the overall power of the weapon. As the size of a gun increases, so does the power of the gun and/or the size of the bullet. Most weapons are built to handle only one type of bullet, or cartridge. The barrel of a gun, therefore, is cut or molded to a specific diameter to allow only one size of bullet to pass through it. This is known as the *caliber* of a gun and is measured as either the inner diameter of a gun barrel or the overall diameter of a bullet. Historically, hundreds of different calibers of weapons have been produced, but today, there are standard calibers for pistols and rifles. The calibers of modern weapons are expressed either as a fraction of an inch (such as .38 or .45 caliber) or in millimeters (such as a 9mm handgun or a 50 caliber rifle).

HANDGUNS

A *handgun*, or pistol, is held and fired in one hand and falls into one of two major types: *revolvers* or *semi-automatics*. (The word *pistol* may have come from the name of an Italian town, Pistoia, where handguns were made in the 15th and 16th centuries. Alternatively, some believe it came from the Russian *pischal*, a word for a 15th century matchlock gun.) A pistol consists of three major components: the *frame* (the backbone to which all other parts are attached), the *barrel* (the metal tube through which the bullet passes), and the *action* (the group of moving parts used to load, fire, and unload the weapon). Although both revolvers and semi-automatics have these three components, some of the parts within these components and the ways in which they function are different.

Revolvers are one type of handgun that consist of a rotating cylinder that holds the cartridges. The cylinder contains hollowed-out spaces for individual

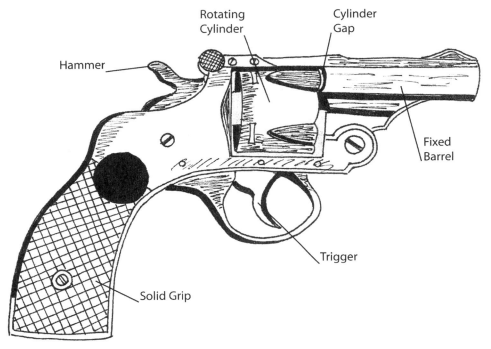

Figure 8-1 Features of a revolver

cartridges—usually six, and hence the name "six-shooter." The action of the trigger causes the cylinder to rotate and fire a bullet, while bringing the next cartridge in front of the gun's firing pin in a sequential manner. The cartridge casings remain in the cylinder after the bullets are fired until they are removed (see Figure 8-1).

Semi-automatics are handguns that are loaded using a magazine, which is sometimes referred to as a "clip." The *magazine* is a spring-loaded storage device that usually inserts into the grip of the weapon and holds a stack of cartridges. A magazine can hold anywhere from 6 to 15 or more rounds of ammunition. All semi-automatics have a *slide*. The slide is always held forward on the gun by a powerful spring. After the semi-automatic is fired, the slide is pushed to the rear by the explosion of the round, ejecting the empty cartridge case and usually cocking the pistol in the process. The cartridges located in the magazine are forced upward by the magazine spring into the path of the slide. When the slide moves forward, it picks up and pushes another round into the chamber. In all semi-automatics, however, the first cartridge must be manually cycled into the firing chamber by retracting the slide and then releasing it (see Figure 8-2).

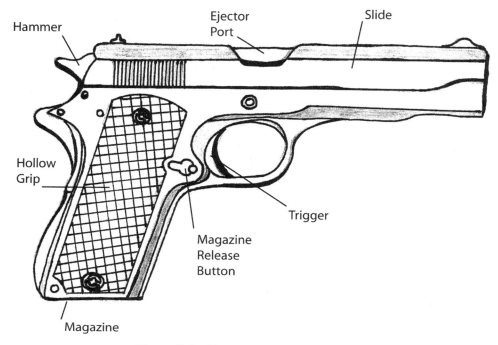

Figure 8-2 Features of a semi-automatic

A *machine pistol*, also known as a *submachine gun*, is a third type of "handgun" weapon that still fires pistol-style ammunition, but operates in a semi-automatic or fully automatic mode. Sometimes, such machine pistols are handheld, like the Israeli-made Uzi, or they can be larger and fired from the shoulder as well, like an MP5 from Heckler and Koch, or a "Tommy gun." Machine pistols are also different from semi-automatic pistols in that the action is usually based on a gas-operated recoil, instead of a slide as seen in a semi-automatic pistol.

RIFLES

A *rifle* is a gun with a long barrel, inside of which there are spiral grooves cut into the metal. These grooves are known as *rifling*, and this is where the rifle gets its name. Most modern guns, however, including handguns, now have rifling in their barrels as well. These rifling grooves help impart spin on a bullet to allow it to fly straighter, much like the way a football is thrown in a spiral fashion. Because of the longer barrel in a rifle, though, they are often used for shooting longer distances or when a larger projectile is needed. A rifle cartridge is much larger than a handgun cartridge, although the rifle bullet is the same size or even smaller than some handgun bullets. Rifles can be semi-automatic or fully automatic, and can be fed ammunition in a variety of ways, using tubes, clips, or magazines. Because of the more powerful nature of a rifle round, the weapon is usually braced against the shoulder and controlled with two hands.

SHOTGUNS

A *shotgun* is a type of weapon that fires cartridge cases containing pellets, known as *shot*, instead of bullets. Shotgun barrels are also much wider than other gun barrels, and are smooth (they do not contain rifling). Because shotgun pellets spread out into a circular pattern, shotguns don't require precise aiming. Also, because the barrel of a shotgun is wide and smooth, there are various sizes of shot that can be used in a shotgun, which allows more flexibility as a weapon. Whereas most guns are measured by their caliber, shotguns are measured by their *gauge*. A shotgun's gauge is equal to the number of lead balls of a diameter equal to the interior diameter of the shotgun barrel that could be made from a pound of lead. So, for example, if the shotgun's barrel is 1 inch in interior diameter, the number of lead balls 1 inch in diameter that equal 1 pound would determine the gauge. If, for instance, 12 of these 1-inch-diameter balls equal 1 pound, then the shotgun is a 12-gauge. Therefore, as the gauge of the shotgun increases, the barrel diameter decreases, and as the gauge decreases, the barrel diameter gets larger. Commonly, most shotguns are 12-gauge shotguns, although there are also 10-, 16-, 20- and 28-gauge models available.

The Components of a Firearm

The most important aspect of all firearms is the way in which they load, fire, and unload. This is known as the *action* of the weapon. The action is made up of several parts. The *chamber* of the action is where the ammunition is inserted when the round is stripped off the magazine, tube, or clip. The chamber is drilled to perfectly fit the cartridge, and is usually set directly into the end of the barrel of the gun, and therefore does not move. The *breech* is the rear part of the action, and moves forward to push the ammunition into the chamber and then locks against the chamber until the round is fired. Once the round is fired, the breech moves backward, pulling out the spent cartridge and ejecting it from the weapon. The breech also contains the *firing pin* mechanism. The firing pin is a spring-loaded steel tip that strikes the ammunition, causing the weapon to fire.

AMMUNITION

Most people refer to ammunition as "bullets," but this is incorrect. A "bullet" is only a part of the ammunition, specifically the part that is discharged by the gun when it is fired. Bullets, before they are fired out of guns, are housed in *cartridges*. A cartridge is therefore made up of the bullet, the casing (or "shell"), gunpowder, and the primer charge. A bullet may be made up entirely of lead, or it may be covered partially or completely by a piece of metal. When a bullet is partially or fully covered by metal (usually copper or steel), this is known as *jacketing*, hence the term "full-metal jacket."

The bullet is firmly set in the end of a cartridge, and it must be set deep enough and tight enough to keep the gunpowder dry and to keep the bullet from falling out. The gunpowder is packed tightly inside the cartridge, and is encased by the cartridge case and the bullet in an intact live round. Sometimes, the cartridge case encases the gunpowder and projectile completely, like in a shotgun shell. While most handgun and rifle ammunition is made using metal cartridge cases, shotgun shells are plastic and folded closed at the top. Inside the shotgun shell, the shot and the gunpowder are separated by *wadding*, which is either a plastic plug or a layer of cardboard.

Regardless of the cartridge type, at the rear of every cartridge is a small button called a *primer cap*, which contains the *primer*. The primer is a high-explosive that can be detonated by a quick shock, like getting hit with a fast-moving firing pin. When the firing pin in a gun contacts the primer cap on a cartridge, the primer explodes and sets off a chain reaction that causes the gunpowder to ignite. As the gunpowder quickly burns, the gases that are produced build up until they are strong enough to push the bullet out of the end of the cartridge and into the barrel of the gun. As the bullet travels down the barrel of the gun, the gases push it along until it exits the barrel and continues to fly through the air (see Figure 8-3).

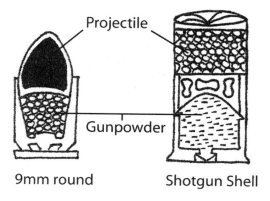

(Note: Not to scale)

Figure 8-3   Cutaway comparison of a 9mm round and a shotgun shell

MECHANICS OF A FIREARM

In some guns, the gases produced from the explosion of the ammunition are also rerouted in the barrel and are used to cycle the action of the weapon to chamber and fire another round. Guns of this nature are known as *gas-operated* weapons, such as the AK-47 and the M16 assault rifles. Other weapons expel all of the gas out of the barrel, and instead rely on the *recoil* of the explosion to push back the breech of the action, ejecting the spent round and chambering a new one. The recoil is the effect of firing ammunition in a gun. Since the chamber is a solid piece of metal and the barrel is hollow, the forces of the explosion need to expand and move down the barrel, pushing the bullet out of the way. The explosion also expends energy rearward at the same time, and in guns that use this recoil, the force of the explosion pushes back the breech, which ejects the spent round and chambers a new one as it moves forward. Guns that use this type of action are usually semi-automatic handguns. There are also other weapons that do not use any of the forces of the explosion to cycle the action, but instead rely on the person firing the weapon to cycle the action. These types of weapons are known as *bolt-action* weapons, and can usually be found in hunting rifles.

MECHANICS OF A REVOLVER

While most guns have the same parts and mechanics to their actions, revolvers are different. In a revolver, the breech is a solid, nonmoving part of the gun. The "chamber" is actually the cylinder that houses the ammunition and spins around. Unlike most guns, there is a gap between the front of the cylinder and the rear of the barrel, as well as a gap between the rear of the ammunition and the breech. As the cylinder rotates into position, the firing pin (which is attached to the *hammer* of the gun) falls forward, striking the primer cap and firing the weapon. As the revolver is fired, the bullet enters the rear of the barrel at the same time the rear of the cartridge is pushed into the breech face of the revolver. Pulling the trigger again causes the cylinder to rotate into the next position. As the cylinder is rotating the next round into position, the hammer is also moving backward. If the trigger is pulled all of the way, the cylinder will lock into position just before the hammer falls, firing the next round. Note that, unlike other firearms, the breech does not serve any purpose in unloading the revolver. A revolver retains all of the spent cartridge cases until the cylinder is swung out and the cartridge cases are manually ejected.

Forensic Examination of Bullets

When a gun is discharged, two items are left behind that can identify that gun. One of them, as previously discussed, is the cartridge casing. The other is the bullet. To understand how a bullet can be used to identify a weapon, it's important to examine the barrel of a gun. All modern gun barrels, with the exception of shotguns, are rifled. *Rifling* is the process in which spiral grooves are cut into the length of the barrel. These grooves help the bullet spin so that it is stable in flight.

Rifling is produced in several different ways—it can be made by dragging a sharp hook through the barrel, or by pushing, hammering, or pulling a rotating cutting bit down the barrel. As the metal inside the barrel is scraped or cut away, *grooves* form. The original metal that is left untouched between each groove is known as a *land*. The lands and grooves therefore make up the rifling of the barrel. The diameter of the barrel, and therefore the caliber of the weapon, is the diameter between opposite lands.

Another important feature of rifling is the *pitch*. The pitch is the "steepness" or the angle of the grooves as they are cut into the barrel. The pitch of a barrel is usually expressed in what the length of the barrel would need to be for the rifling to travel one inch around the circumference of the barrel. Rifling usually has a fairly shallow pitch so that the bullet can still quickly move down the barrel yet get enough spin from the lands and grooves for a straight flight.

Finally, the direction of the rifling is important as well, and this is known as the *twist*. The twist of a barrel can be one of only two possibilities: right-hand twist or left-hand twist (see Figure 8-4).

All of this rifling information is important because it can serve to identify the make and model of the gun. For example, if just the barrel of a broken gun was recovered from a crime scene, the examination of the lands and grooves inside the barrel might show a left-hand twist of six lands and six grooves. The pitch of the rifling, when measured, might show a pitch of 1 inch in 16 inches. If the diameter between opposite lands is measured and found to be 0.45 inch, then you know it is a .45-caliber barrel, and based on the rifling, it possibly could be part of a Colt

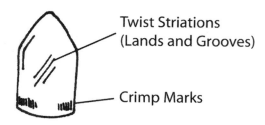

Figure 8-4   Possible toolmarks on a bullet

M1911A1 pistol. A forensic firearms examiner would have access to databases and reference books with information on all types of weapons, including rifling data. However, it would be a very rare occurrence to find only the barrel of a gun at a crime scene. The rifling data that is in databases and books is mainly used when examining bullets.

When a gun is fired, the bullet is forced out of the cartridge and into the end of the barrel at high speed and under extreme pressures and temperatures. Because bullets are made of soft lead and other metals, they quickly expand to fill in all of the space they can inside of the barrel, as they are being pushed from behind. As the bullet expands, the lands press into the soft surface of the bullet as the bullet forces itself into the grooves of the barrel. Once this happens, the bullet has no choice but to start rotating to follow the paths of the lands and grooves down the length of the barrel. When the bullet exits the barrel, then, it is spinning in the same direction as the twist of the lands and grooves in the barrel. Also, the impressions of the lands are left behind in the soft surface of the bullet. If the bullet is later recovered, the impressions in the surface of the bullet, as well as the right or left direction of spin, can be established and linked back to the firearms databases and books to figure out the make and model of the weapon that fired the round. Just as importantly, though, these same markings on the bullet surface can help individualize a single weapon as well.

Because rifling is produced using mechanical methods, no two guns are the same. As the rifling is cut into the barrel, minor imperfections can occur that leave behind individual features in each barrel. Sometimes, the barrel might be a hair thicker in one section, which causes the land in that one region to cut into the bullet just a little deeper than the rest. Sometimes, the rifling cutter itself might stall, skip, or jump, causing an imperfection in the rifling pattern of the weapon. Most commonly, the action of cutting the metal inside the barrel leaves sharp barbs on the edges of the lands or in the bottom of the grooves. These barbs can striate the bullets in unique ways, and therefore leave an individual mark upon the bullet that relates directly back to that gun.

The above aspects of forensic bullet analysis are all well and good, but for any of the examinations to be done, the bullet itself must first be recovered. The recovery of a bullet poses a fundamental paradox for a forensic firearms examiner—how can an intact bullet be recovered from a crime scene that has only the markings from the barrel, and nothing else? The answer is that it cannot. In order for a bullet to stop, it must hit something. At the scene of a shooting, that might be an object or a person. Depending on the circumstances of the shooting, the bullet might have bounced off other objects along the way, and therefore ricocheted into the final target. If the bullet can be recovered, then, it not only will contain the markings from the lands and grooves of the gun, but also will be deformed by whatever it hit, scraped against, or bounced off. It is the job of the forensic firearms examiner to look past the deformity of the bullet and establish the original markings to the best of their ability. This can be difficult. The best place usually to recover intact bullets at a crime scene is, unfortunately, a body.

When a bullet comes into contact with a person, the energy of the bullet can quickly dissipate into the body, which slows the bullet down. If a small enough and light enough round is used, the bullet may stay in the body. However, the bullet will still have markings on it from the interaction with the body. The nose of the bullet will probably deform from hitting the clothing or skin of the person, and if the bullet glances off a bone, it can be scraped or marked in a damaging way.

The opposite is true if the bullet exits the body or hits a harder, denser object. Bullets that exit the body can begin to tumble in midair. When they finally come to rest, the side or the rear of the tumbling bullet might make contact with an object. If a bullet hits a hard, dense object like a steel plate or a hard rock, the bullet may completely break apart and be unrecoverable.

If and when the bullet is recovered, it still must be compared against a known bullet to match it with a single firearm. While the rifling data from a bullet can narrow down the make, model, and caliber of a firearm, there still might be thousands of weapons out there of that make, model, and caliber. If a suspect weapon is recovered at a crime scene, a test firing must again be performed in the lab to produce an exemplar bullet that can be compared to the unknown evidence bullet. This test firing is done in a controlled way, and the bullet is fired into a *water tank* or cotton wadding. When a bullet is fired into water, the water quickly dissipates the energy of the bullet and slows the bullet down. Because the water is much less dense than the metal of the bullet, the striations on the surface of the bullet remain pristine. The exemplar bullet can be fished out of the tank and compared under a compound microscope (see the upcoming sidebar "The Comparison Microscope Sidebar") alongside a suspect bullet to see if the individual markings are a match. If they are, the chances are good that the weapon that produced the exemplar bullet is the same that was used in the crime.

Forensic Examination of Cartridge Casings

Empty cartridge casings are the most commonly encountered evidence at the scene of a shooting. The discovery of a spent cartridge casing at the scene can aid the forensic investigator in narrowing down the type of weapon used. The lack of cartridge cases at a shooting scene may indicate that the weapon used was a revolver, but the shooter could have also picked up the spent cartridge cases, so although it is important to make a note of the cartridge casings, nothing should be assumed by their presence or absence.

The most important aspect of collecting a cartridge casing is to make sure not to damage or scratch the cartridge in any way. The spent cartridge case, ejected from whatever type of gun fired it, is covered with informative toolmarks that can not only identify the weapon, but individualize it as well. If the shooter handled the rounds with his or her bare hands, the cartridge cases may also contain fingerprints and/or DNA. After the location and the position of the cartridge cases at a crime scene have been documented, the cartridge cases should be collected with soft-nosed tweezers and sealed individually in plastic bags.

First and foremost, the caliber of the weapon that was used in the shooting can be established by the cartridge case. Usually, the caliber is stamped on the rear face of the cartridge case, on the metal rim surrounding the primer cap, as well as the ammunition manufacturer's name or insignia. If there is no stamped information, the diameter of the cartridge case itself can be measured to determine the caliber of the bullet.

Secondly, the toolmarks that are formed by loading, firing, and unloading the weapon can be used to individualize the cartridge case back to one weapon.

Before a gun can fire, the cartridge must be fed into the chamber of the weapon. As this happens, the soft metal of the cartridge scrapes against the hard metal of the chamber, which produces *striations* in the cartridge metal. Striations occur when one object drags across another object and leaves a mark or trail. As the breech closes and locks, the rear of the cartridge is pressed against the face of the breech, and any imperfections in the breech face can be impressed into the metal of the cartridge as well. When the trigger is pulled, a series of levers release the spring-loaded firing pin and the point impacts the primer cup on the rear of the ammunition. This point presses into the metal, crushing and igniting the primer and leaving an indentation in the surface of the primer cap. Sometimes, the shape of the firing-pin impression can be used to ascertain the manufacturer of the gun, most notably the Glock, with its tear-drop-shaped firing pin.

When the gun is fired, the cartridge case bears the brunt of the explosion. Because the cartridge case is made of thin metal, this metal expands and flexes during the combustion of the gunpowder. This expansion of the cartridge case forces the soft metal to further press into the walls of the chamber and strongly push against the breech face. As the spent cartridge case is ejected from the chamber, the breech uses an *extractor arm* to grasp the end of the cartridge and pull it out of the chamber. As this occurs, both the extractor arm and the chamber walls again leave marks on the soft metal of the cartridge. Finally, as the empty cartridge case is traveling rearward, it impacts a small ramp that causes the cartridge case to pop up and out of the open action, and eject out of the weapon. The ejector ramp also leaves a mark on the cartridge case. These marks are always made every time the gun is fired, and are unique to each weapon. However, the minor defects that exist in the chamber and breech of a gun can change over time, because just as these defects leave marks on the cartridge metal, they also wear down from the friction and heat that is produced when a gun is fired (see Figure 8-5).

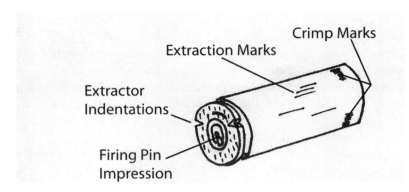

Figure 8-5 Possible toolmarks on a cartridge case

To compare the spent cartridge cases from a crime scene to a weapon, that weapon must be recovered within a relatively reasonable amount of time. Because the imperfections that cause the individual markings can change over time, it might be unreasonable to expect to match a gun against an unknown cartridge case if several years have passed and the gun was in use during that time. However, if a suspected weapon is recovered from a crime, a *test firing* in a forensic laboratory will produce an *exemplar* cartridge casing that can be used to compare against the unknown cartridge casings from the scene of the crime. When doing a test firing in a lab, it is ideal to use the same or similar type of ammunition that was used during the shooting, so that the comparison of the known and unknown cartridge cases is as close as possible. While the minutiae of each cartridge casing can be determined by direct examination, the quickest way to establish a possible match is with a comparison microscope (see the sidebar "The Comparison Microscope").

The Comparison Microscope

The comparison microscope is an invaluable tool for a forensic firearms examiner. The basic concept is that an *optical bridge* joins together two separate microscopes in one view, so that one half of the view is from one of the microscopes and the other half is from the second. Therefore, if similar items, such as two pennies, are on the two stages and under the same magnification, the person using the comparison scope would see the left half of the penny from the left microscope and the right half of the penny from the right microscope all at once. If the pennies were *exact* duplicates of each other, down to the scratches on the surface and the coloration, there would appear to be no difference between the right half and left half of the image in the comparison microscope. But, if the pennies were different, as all pennies are, the scratches or coloration on the left penny would not line up with the

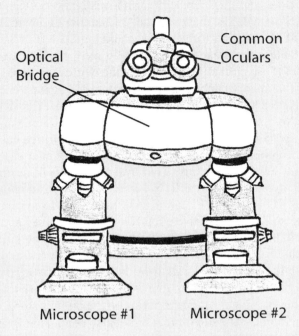

Microscope #1   Microscope #2

scratches or coloration on the right penny, and they would look like two different halves of the penny. By using this microscope setup, two objects that are suspected of having a similar origin can be compared side by side and the markings can be matched up.

When examining unknown and exemplar cartridge casings, both the breech face marks on the rear of the cartridge case and the striations on the sides of the cartridge cases must be examined and matched up using the comparison microscope. Also, the extractor marks and the ejector ramp marks will be in the relatively same positions on each cartridge case, and these can be compared from cartridge case to cartridge case as well. If the comparison analysis shows that the markings on the unknown cartridge case do not match up with the markings on the exemplar cartridge case, then the gun is ruled out as having been used in the shooting. However, if all of the marks, striations, and minutiae between the known and unknown cartridge cases match up, it can be determined that the weapon that was recovered is the same weapon that produced the unknown cartridge cases from the crime scene.

Forensic Firearm Databases

There are times during forensic firearm investigations when the actual weapon may never be recovered. The make, model, and caliber of the unknown weapon might be determined from the cartridge casings and bullets left behind, but what good does that do? Unless the weapon is extremely rare, chances are that there are several thousand of those particular weapons in the world. The cartridge casings and bullets from the unknown weapon cannot be compared against exemplar samples from a known weapon, so no individualization can be done, right? Wrong!

Thanks to a nationwide system of databases, the forensic information from one crime can be compared against the bullets and cartridge cases from another crime in another region of the country, as well as the exemplar cartridge cases and bullets used in the lab comparisons. This system is known as the National Integrated Ballistics Information Network (NIBIN), and it is run and maintained by the Bureau of Alcohol, Tobacco, and Firearms (ATF). The NIBIN system utilizes a very important tool to collect and archive data from cartridge cases and bullets, the Integrated Ballistics Identification System (IBIS). IBIS is a special system of digital cameras and computers that connect to a compound microscope and allows digital images of the individualizing markings from bullets and cartridge cases to be stored and compared. For example, if a bullet is documented on IBIS, the file for that bullet includes digital images of the striations on the bullet's surface as well as other individualizing marks. The file for that bullet also contains information on the probable caliber of the bullet, and perhaps even the possible manufacturer of the bullet. This IBIS information is then uploaded to NIBIN, where it is stored in a master database and compared to all of the other entries for known and unknown bullets from around the country. If the computer finds a possible match, the NIBIN user from the original firearms lab is notified as well as the other lab that uploaded the matching information. If the NIBIN match links two unknown bullets, then they still might not individualize to a particular weapon, but if the case details are compared, they may show a link between the two crimes, or establish a timeline of where the gun was during the times of those crimes. If the NIBIN match links the unknown bullet with an exemplar bullet from another lab, then the individual weapon can be established, and if the person possessing that weapon is known, they can be linked to the crime.

The NIBIN system has proven invaluable to law enforcement in regards to illegal gun trafficking and smuggling. Sometimes, when a weapon is used in a crime in one area of the country, the owner of that weapon sells or trades it to a gun runner. The gun runner then transports the suspect weapon to another region of the country and again sells or trades it. If gun runners are caught, their weapons are taken to a forensic firearms lab and test firings are done on all of them. The exemplar information that is loaded into NIBIN can then link these illegally transported guns to

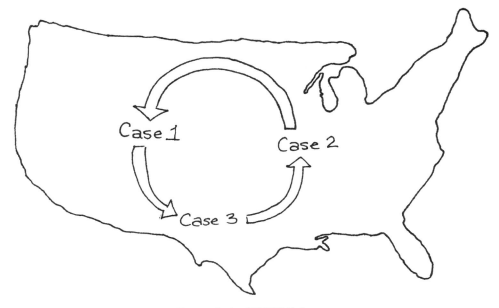

Figure 8-6   NIBIN link map

crimes in other areas of the country, and establish proof not only of the smuggling but of the original shooting crime in the other city (see Figure 8-6).

Other Specialties of Forensic Firearms Investigation

The forensic investigation into crimes involving firearms is not limited to just the analysis of the weapon, the bullets, or the cartridge cases. There are several other types of evidence that can be present at the scene of a shooting. Therefore, key questions about the position of the shooter and the victim, the distance between the shooter and the target, the number and order of shots fired, and even the trajectory of the bullets can be answered by properly collecting and analyzing the evidence.

GUNSHOT RESIDUE

When a weapon is fired, the primer in the cartridge is quickly shocked in such a way to cause it to explode. This explosion causes the gunpowder inside the cartridge case to ignite and burn. As the gunpowder burns, it gives off gases which quickly expand and force themselves out the path of least resistance. Since the bullet is firmly wedged but not permanently connected to the cartridge, it is forced out by the expanding gases and into the barrel. The gases and still-burning gunpowder also move into the barrel behind the bullet and expand further and further as the bullet moves down the barrel. As the bullet exits the barrel, so do the gases and remaining gunpowder from the explosion. Some of these gases are the result of the burning gunpowder. Modern gunpowder is commonly made of a mixture of potassium nitrate, charcoal, and sulfur. As gunpowder burns, chemical compounds called *nitrites* are produced, along with organic chemical residues. Also present in the gases of a gunshot are the remains of the primer explosion. Modern primers are made of a combination of lead styphnate, antimony sulfide, and barium nitrate. The primer gases therefore contain traces of lead, antimony, and barium. When any weapon is fired, the gases emitted from the firearm have a tendency to settle on the hands and clothing of the person firing the weapon. In some weapons, such as a revolver with the gap between the cylinder and the barrel, the amount of these gases emanating from the weapon is even greater. As the gases settle on the hands and clothing of the shooter, they become *gunshot residue (GSR)*.

Gunshot residue contains all of the chemicals that are left over from the burning of the gunpowder and primer. However, the presence of nitrites is not really a unique identifier of GSR, because there are thousands of products around that also contain nitrites. Every time you handle a newspaper or pick up a plastic cup, you come into contact with nitrites. So, while nitrites are not that useful in GSR analysis, the presence of lead, antimony, and barium most definitely is. If a person is suspected of having recently fired a gun, their hands and clothing can be sampled and tested. To collect GSR, a dilute concentration of nitric acid is applied to a cotton swab and the swab is brushed over the back of the hands, the fingers, and the webbing of the thumb. The palm is generally not swabbed, because to hold the gun, the palm of the hand is around the grip of the weapon, and therefore not exposed to the gunshot gases. The swabs can then be tested using a variety of instrumental methods to look for lead, antimony, and barium. To conclusively prove that GSR is present, all three of the elements must be found to be present. Also, there is a unique feature of GSR that can also help to identify it. As the gases from the gunshot cool and settle on the hands and clothing of the shooter, the superheated chemicals harden into microscopic spheres. These spheres can be seen on a swab using a *scanning electron microscope (SEM)*.

A scanning electron microscope is a special type of microscope that uses electrons to "trace" over the contours of a surface. Because electrons are so small, it is possible to achieve incredibly high magnification and detail using an SEM. The spheres of gunshot residue are so small that they cannot be seen under a normal microscope, but by using an SEM, they can be visualized in 3-D. At the same time, an SEM can also be equipped with an *energy-dispersive x-ray spectrometer (EDX)*. When the electrons from the SEM come into contact with the GSR spheres, some electrons are altered by the chemistry of the specific elements present on the swab. The EDX is a detector that can specifically look at the way that the electrons are altered by the material that is being scanned. Every element causes a unique change in the nature of the electron, and therefore the way the electron is altered is unique to specific elements. The EDX is therefore capable of screening the GSR spheres for the presence of lead, antimony, and barium. If the three GSR elements are found and the spheres are present, then a forensic firearms examiner can determine that the swabs must have come from a person who recently fired a gun.

RECONSTRUCTION OF A SHOOTING

Sometimes it is necessary for a forensic firearms examiner to go to the crime scene to reconstruct the shooting. This might be done to help find a missing bullet, to establish the position of a shooter, or to determine how far the shooter was from the target. Because a bullet travels a fairly straight path over a short distance, the crime scene can be reconstructed simply by using string, or it can be reconstructed in a high-tech way by using lasers and computers. Regardless of the method, the ultimate goals of the reconstruction are the same each time, and a good forensic scientist will use the scientific method to find all of the answers.

One of the common questions at a shooting scene is the location of the shooter. If the location is known, it can help prove or disprove a suspect's account of the crime. One of the ways this can be established is by using the bullet holes at a scene. Because a bullet can travel in only one direction, and that direction is away from the shooter, the bullet holes at a scene can help establish the location of the shooter.

Sometimes, though, the holes might be found in a surface that has moved, and it may be unclear on which side the bullet entered and on which side it exited. To figure this out, a firearms examiner can use *sodium rhodizonate*, an orange liquid that reacts with lead to form a purple color. When a lead bullet passes through a surface, the edges of the surface are pushed in and away from the bullet as it moves through. The rear side of the surface, where the bullet exits, is also pushed away from the bullet, but does not come into contact with the bullet at all. The front side of the surface is therefore the only side that contacts the bullet. This is known as the *bullet wipe*, and is simply a smear of lead. If sodium rhodizonate is sprayed on each side of the surface, only the entrance side with the bullet wipe will react to form the purple coloration, while the exit side will not. Therefore, sodium rhodizonate can help in establishing the direction of the bullet at a crime scene (see Figure 8-7).

Another important factor to establish at a shooting is the distance from the shooter to the target. This is important because establishing the true distance can prove or disprove a suspect's account of the crime. Distance determination can be

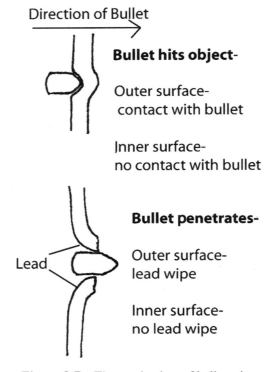

Direction of Bullet

Bullet hits object-

Outer surface-
contact with bullet

Inner surface-
no contact with bullet

Bullet penetrates-

Lead

Outer surface-
lead wipe

Inner surface-
no lead wipe

Figure 8-7   The mechanism of bullet wipe

done in two ways. Both methods rely on the *stippling pattern* that is present or absent around the entrance wound or bullet hole. Stippling is the result of the unburnt or still-burning particles of gunpowder hitting the target at high velocity. As a gun is fired, these particles are ejected from the barrel just as fast as a bullet but not as accurately. When they come into contact with a surface, they appear as small pockmarks ("punctuate abrasions") around the bullet hole or entrance wound. However, for stippling to occur, the distance from the shooter to the target must be small, and the overall distance depends on the caliber and the type of weapon. Generally, a pistol can cause stippling within 2 to 3 feet, whereas a rifle can cause stippling from within 5 to 7 feet. The closer the shooter is to the target, the "tighter" the stippling pattern will be around the entrance wound or bullet hole. The further away the shooter is from the target, the more spread out the stippling pattern will be. So, the first factor that can be looked at for distance determination is the presence of stippling and, if present, how much area the stippling covers on the target (see Figure 8-8).

Sometimes, the stippling might be very faint or hard to observe because of the nature of the surface that was shot. Because the stippling is the result of the high-velocity contact with the burning or unburnt gunpowder, the *Greiss test* can also be

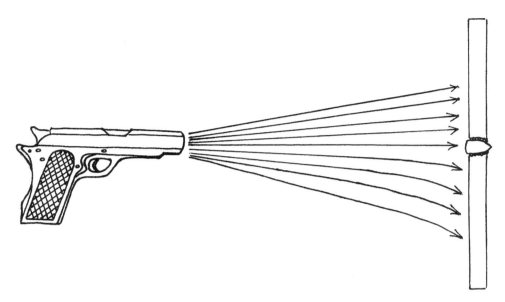

Figure 8-8   Stippling

used to screen for the presence of nitrites. Greiss reagent turns orange in the presence of nitrites and heat, and if stippling is suspected, the area around the entrance wound can be tested with a piece of paper soaked with Greiss reagent. If orange specks appear after the paper is heated, then nitrites may be present, which are indicative of stippling. The size of the area that reacts with the Greiss reagent is therefore the same as the area that is covered by the stippling, and the distance can again be estimated.

The overall determination of distance is important because it puts the physical relationship of the shooter and victim into perspective. For example, suppose that the suspect in a shooting homicide claims that he shot the victim with a 9mm semi-automatic in self defense in a hand-to-hand struggle. If the body is then checked for stippling and none is present, and the Greiss test is negative around the entrance wound, then the suspect's story of a close-distance shot does not add up. For there to be no stippling and no GSR on the victim's body, the victim must have been more than 3 feet away from the suspect, and therefore was not fighting hand-to-hand.

SERIAL NUMBER RESTORATION

Another important aspect of forensic firearm examination is the restoration of serial numbers on suspect firearms. When a firearm is produced by a company, a serial number is stamped into the metal of the gun, usually on the barrel or on the action of the weapon. This serial number is unique to that one gun, and if the gun's information is ever entered into NIBIN, the serial number will be used to individualize that one gun as well. If the gun is used in an illegal way, the owner might then opt to file down the serial number so that it cannot be read. While this obliterates the number itself, the underlying metal still contains the impressions of the original numbers. When metal is stamped, the metal directly under the stamp becomes more compressed than the surrounding metal. The compressed metal under the stamped area is under a permanent strain and can be altered, whereas the unstrained metal cannot. To restore a filed away serial number, a firearms examiner polishes the metal where the serial number was filed away, and then adds a thin layer of *Fry's reagent* to the metal. Fry's reagent is an *etchant*, which means that it eats away metal, but Fry's reagent is unique in that it dissolves only the metal that is under permanent strain, and not the uncompressed surrounding metal. Because the surrounding metal is untouched, the compressed metal is eaten away and shows the original serial number. There are also various other chemical, electrochemical, and magnetic methods of serial number restoration, and they all rely on the principle of altering the strained metal where the original serial number used to be.

Toolmarks

A large part of forensic firearm examination is based on the principles of forensic toolmark analysis. Toolmarks are encountered in a forensic setting when a denser object is used on a less-dense object in the course of a crime. This can be something simple, like a crowbar that is used to pry open a wooden door, or a screwdriver that is used to smash a lock. When one object is used on another object, an impression of each object is left behind on each object (which is Locard's Exchange Principle!), and this is known as *impression evidence*. In the case of the crowbar and the wood door, the door frame will contain impressions from the crowbar bit, as well as the handle and shaft as it all pressed into the wood. Likewise, the crowbar will pick up flakes of paint or splinters of wood from the door. The impressions of the crowbar in the wood, however, are unique to that crowbar, and if the impressions are cast, a three-dimensional image of the tool that was used to create the impression can be obtained.

There are several different casting materials available for forensic purposes, and they vary depending on the surface that is being cast and the location.

Impression evidence can also be something as simple as a shoeprint or tire tread in mud, but these too can be cast to show what the original object looked like. The importance of the casts is to collect both the rough dimensions of the tool and the individualizing marks that can single out that tool from all others. For example, if the crowbar used on the wood door had a large gash in the metal, and that was pressed into the wood frame, then the impression of the gash is a distinguishing feature of the impression, and can be matched directly back to that one crowbar.

Another type of toolmark that is encountered at crime scenes is striations. Striations are formed when one object is dragged across another, like a bullet across the rifling of a gun's barrel. Striations can also occur on other objects. For example, if a car was involved in a robbery and was observed to sideswipe a fire hydrant as it left the scene, the fire hydrant could have left striations on the side of the car that are unique to that fire hydrant. Likewise, the car probably left paint or metal on the hydrant that is unique to that one car.

One thing to keep in mind about striations, though, is that striations made by a tool can differ based on the angle of that tool. For example, imagine the prongs of a fork dragging though the icing on a cake. If the fork is held so that the prongs are all straight across, the lines that are created are the same distances apart as the prongs. But, if the fork is turned a little bit sideways and dragged through the icing again, the lines are now closer together. Even though the two patterns are different, the same tool was used to create both (see Figure 8-9). A forensic examiner must keep this in mind when analyzing all types of toolmark evidence.

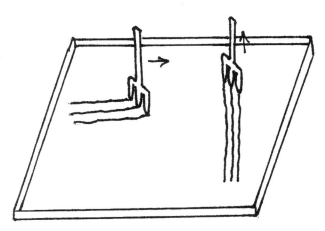

Figure 8-9   Different striations from the same object

Quiz

Refer to the text in this chapter if necessary. Answers are located in the back of the book.

1. The features of a rifle barrel that cut into a bullet and impart spin are:
    (a) Slots
    (b) Slides
    (c) Ramps
    (d) Lands

2. Which of the following is not a component of a semi-automatic pistol?
    (a) Ejector rod
    (b) Firing pin
    (c) Slide
    (d) Grip

3. Which of the following is not a component of a rifle?
    (a) Action
    (b) Cylinder
    (c) Breech
    (d) Magazine

4. Toolmarks are commonly encountered as:
    (a) Impressions
    (b) Striations
    (c) Lines
    (d) Both A and B

5. Where should a firearms examiner engrave their initials on a questioned shell casing?
    (a) Primer cap
    (b) Outside of the shell
    (c) Rear face of the shell
    (d) Inside lip of the shell

6. The gauge of a shotgun is determined by:

    (a) The exterior diameter of the barrel

    (b) The interior diameter of the barrel

    (c) The number of lead balls that would equal a pound

    (d) A combination of B and C

7. Rifling of a barrel is produced by:

    (a) Cutting away metal to form grooves

    (b) Adding layers to metal to form lands

    (c) Pouring the metal into a mold

    (d) Both A and B

8. A screwdriver used to pick a lock is scratched by the lock's tumblers and leaves pieces of metal behind. This is an example of:

    (a) Pattern evidence

    (b) Toolmark evidence

    (c) Locard's Exchange Principle

    (d) All of the above

9. The elements that are determinative of gunshot residue are:

    (a) Lead, arsenic, boron

    (b) Lead, antimony, copper

    (c) Antimony, barium, arsenic

    (d) Lead, antimony, barium

10. The color-changing chemical that detects the presence of lead bullet wipe is:

    (a) Dille-Koppanyi reagent

    (b) Frye's reagent

    (c) Sodium rhodizonate

    (d) Kastle-Meyer reagent

CHAPTER 9

Drugs and Toxicology

Toxicology is the study of *toxins*, or poisons. A forensic scientist that works in a toxicology laboratory is known as a *toxicologist*. Historically, toxicology focused on the detection and measurement of classic poisons like strychnine, arsenic, thallium, lead, and mercury. Although these poisons are still used to commit crimes today, the main focus of modern toxicology has shifted away from the detection of such "classic" poisons and tends to focus more on the detection, isolation, and quantitative analysis of drugs.

Drug abuse and drug-related crimes contribute to a large number of cases in the legal system worldwide. Illegal drug sales, the abuse of both illegal and legal (prescription and over-the-counter) drugs, and the diligence of police narcotic units keep today's drug and toxicology labs very busy. Drug-related evidence can come in many forms to a lab. A forensic chemist or criminalist may test a suspected powder in a baggie, or the residue of drugs in a needle. Toxicology specimens are found in body fluids and tissues of a suspected user or victim, and these samples are tested for traces of a drug. Many forensic labs employ both toxicologists and forensic chemists (sometimes called criminalists). Generally, toxicologists examine tissue samples and blood and urine specimens for drugs and toxic substances, while foren-sic chemists test solid dose drug submissions. If the toxicology lab and the drug testing, or *narcotics* lab, are in different agencies, the drug evidence would go to the drug lab and the toxicology evidence would go to the toxicology lab.

When body fluids and tissue samples come into a toxicology lab, the testing that is performed on those items may not test for the actual drugs themselves, but rather for the *metabolites* of the drugs. When the body's chemistry begins to separate sub-stances into smaller parts, this is known as *metabolism*. The process of metabolism creates metabolites, which are the byproducts of substances after the body breaks them down. Drug metabolites are not the only metabolites tested for in a toxicology lab. There are also general metabolites of certain nontoxic substances that are toxic to humans. For example, methanol, which is an alcohol made from wood, is not normally toxic to humans, but if it is consumed, the liver metabolizes the methanol to formaldehyde, a toxic chemical that can cause blindness or death.

With other types of drug evidence, such as containers of illicit drugs or used needles, a drug chemist will actually test for the presence of the drug itself. If a known drug is discovered, an attempt may be made to *quantify* the drug as well. *Quantitative analysis* can basically be thought of as counting, and in toxicology it is a process of weighing or measuring the amount of drugs or narcotics present in a sample. Quantitative analysis is important because drugs, poisons, and other toxins might be *adulterated*, or "cut," with a second chemical, liquid, or powder to make the drug less potent and less expensive. Due to the stringency of the current drug laws in some states, even having 1 percent of a drug in a sample could send the owner of the sample to prison. In some states, the cutting agent weight is used in measuring the aggregate amount of the drug. In certain other states where the laws are different, a person can only be charged with possessing minimum quantities of the actual drug, not the drug plus the cutting agent. However, these minimum quan-tities are extremely small, and many labs still test the aggregate weight of a sample and then determine the percentage of drug that is present by quantitative analysis. For example, a person might be caught with a small baggie of heroin in their pocket. The chemist that tests that baggie might actually find that the heroin is cut with quinine (a white powder with a bitter taste that is commonly mixed with heroin). In fact, if there is so much quinine that there ends up being less than 0.01 grams of total heroin in the baggie, the drug charges might be dropped since there is so little of the actual drug present in the sample. Other common cutting powders include mannitol, lactose, ground-up aspirin, and vitamins. If a drug is in a liquid form, it can be cut with water. In general, a cutting agent is inexpensive and may mimic the chemical properties of the drug that is being cut.

Drugs

Many drugs are encountered on a daily basis in narcotics and toxicology labs. The popularity of certain drugs changes over time, but the effects of such drugs never change, and people who are addicted to them or sell them for a living will always place a burden on the drug chemist and/or toxicologist in the lab. While an entire book could be written on toxicology and drugs, this section provides a brief outline of some of the common controlled substances that are tested for in a toxicology or drug chemistry lab. An important factor to consider when talking about drugs is the *schedule* of that drug, based on the Controlled Substances Act of 1970. Section 812 of this legislation groups all drugs, legal and illegal, into one of five categories:

- **Schedule I** Drugs and substances that have a high potential for abuse and no accepted medical use or value. Examples of drugs in this category are heroin, LSD, ecstasy, and marijuana.
- **Schedule II** Drugs and substances that have a high potential for abuse, which could lead to severe physical or psychological dependence, but that do have an acceptable medical use or value under strict supervision and legal restrictions. Examples of drugs in this category are cocaine, morphine, methamphetamine, and fast-acting barbiturates.
- **Schedule III** Drugs and substances that have less potential for abuse than Schedule I or II drugs and substances, such that abuse of the drug could lead to low or moderate physical dependence or high psychological dependence, but that do have an acceptable medical value. Drugs in this category include low-level opiates like Vicodin and moderate-acting barbiturates.
- **Schedule IV** Drugs and substances that have a low potential for abuse relative to the drugs and substances in schedule III, such that abuse of the drug or substance may lead to limited physical or psychological dependence, but that do have a safe and acceptable medical use or value. Drugs in this category include most synthetic drugs, like Xanax and Valium, and slow-acting barbiturates.
- **Schedule V** Drugs and substances that have a low potential for abuse relative to the drugs or other substances in schedule IV, such that the abuse of the drug or substance may lead to a limited physical or psychological dependency, but that still have a safe and acceptable medical use and value. Most of the drugs in this category are the basic over-the-counter painkillers and cold remedies.

The scheduling of drugs and substances is updated and maintained by the Department of Justice and the Food and Drug Administration. The schedule of a drug might move from one schedule to another based on the current levels of abuse or the creation of a newer drug with less potential for abuse.

COCAINE

Cocaine is a white or off-white powder that is isolated from the leaves of the coca plant. Cocaine is a *stimulant*, which means that it affects the central nervous system of a user to produce a rush (or "high") and gives the user a feeling of energy and euphoria. Cocaine can be snorted, smoked, eaten, or injected *intravenously* (with a needle). The street names of cocaine are coke, snow, blow, powder, and white, among others. Cocaine is a Schedule II drug because of its high risk of physical and psychological dependency, but it is also still used as a topical anesthetic for ear, nose, and throat surgery.

Cocaine went through a period of huge popularity in the 1970s and 1980s, but toward the end of the 1980s until now the use of the powdered from of cocaine has diminished in favor of "crack" cocaine. Crack is cocaine that has been boiled with sodium bicarbonate (baking soda) and cooled to a hard crust. When the crust is broken up, the rocks are known as crack. Crack is the freebase form of cocaine, and it is not water soluble, but crack has a low boiling point, so it can be vaporized with a flame and "smoked." The vapors from a crack rock are more potent than the normal effects of cocaine, and users of crack feel intense rushes. This form of cocaine, however, is more dangerous, and several people each year die from the physical reactions to smoking crack.

NARCOTICS

Narcotics are a class of drugs that include heroin, opium, morphine, and OxyContin. The term "narcotics" is sometimes used generically to describe all types of drugs, and is commonly used in terms such as "narcotics unit," but this is a bit of a misnomer. Not all drugs are narcotics. A narcotic is a type of drug that has a *depressant* effect on the central nervous system, which is to say that it produces a sleepy or dreamlike state in the user. Narcotics are also associated with a high level of addictiveness, and long-term users of narcotics can build up tolerance to the drug's effects.

All naturally occurring narcotics are isolated from the milky sap of an unripened seed pod of the opium poppy. Opium, which is the most natural form of any narcotic, is the cooked, dried sap of the poppy, and can be eaten or smoked. Opium can be purified to isolate a more powerful drug, which is morphine. Morphine exists naturally as an opium alkaloid and is prepared by chemical extraction from opium. If morphine is put through a specific set of chemical reactions, heroin can be synthetically produced. Heroin, therefore, is a chemical derivative of morphine, and heroin is extremely potent.

Several decades ago heroin was exclusively an intravenous drug, which required a user to perform an elaborate "cooking" procedure to get the drug into liquid form each time they wanted to use the drug, or "shoot up." These days, however, the purity of heroin is so high that users can snort the drug to achieve the same effect as an intravenous shot. Heroin is the most common form of any narcotic that is found on the street, and its street names are horse, H, junk, and smack, among others. Chronic abusers of heroin are sometimes referred to as "junkies."

There are other forms of narcotics that are not isolated from the poppy, but are instead synthetically made. One of these drugs is OxyContin, which is a prescription painkiller that is designed by the manufacturer to have a delayed release over certain periods of time, thanks to a coating on the pill. Abusers of this narcotic remove the time-release coating and get the full effect of the drug all at once, which can quickly lead to tolerance of and addiction to the drug. Another commonly abused opiate derivative is Vicodin, which has the same effects as OxyContin but in lower doses.

Because of the different forms that narcotics can be found in, they do not all fall under one simple schedule. Heroin, due to a complete lack of medical value, is considered a Schedule I drug. Other opiate derivatives, such as morphine, opium, and OxyContin, are classified as Schedule II drugs because they have limited medical use. Weaker opiate derivatives, such as Vicodin and codeine, fall under the Schedule III classification because they have a lower possibility of abuse.

AMPHETAMINES

Amphetamines are synthetic stimulants that affect the user to create a heightened sense of awareness and alertness, as well as restlessness and a lack of appetite. The most commonly encountered amphetamine is methamphetamine, commonly known as meth, crystal meth, crank, ice, or simply speed. Amphetamines have become increasingly popular over the last decade due to their low cost and ease of preparation, needing little more than over-the-counter decongestants to act as the base for creating the drug. However, the home-based "meth labs" (also referred to as clandestine labs) where the drugs are made sometimes produce extremely toxic byproducts and have a high risk of explosion. Methamphetamines are extremely addictive and, combined with low cost and high availability, have become the drug of choice for many drug abusers. Currently, the United States classifies methamphetamine as a Schedule II drug, but only because commercially produced methamphetamines have been used in the past for certain medical conditions. When produced in a clandestine lab, homemade methamphetamine is less pure but still extremely dangerous. Because of this, legislation is in place around the country to limit the sale of over-the-counter medications that can be exploited for the creation of methamphetamine, and several addiction-treatment facilities have been created that are specifically aimed toward methamphetamine abusers. Even with these procedures in place, methamphetamine use and abuse still has the potential to reach epidemic proportions.

A close relative of amphetamines, MDMA (3,4-methylenedioxymethamphetamine) is a synthetically produced substance, which is better known as the drug Ecstasy. Ecstasy became popular in the late 1980s and surged in illegal use in the 1990s as a popular "rave drug." The effects of ecstasy are a whole-body feeling of euphoria, as well as having a heightened sense of touch, hearing, taste, smell, and sight. These effects are due to the release of serotonin, a naturally occurring chemical in the brain, which is normally released in small amounts to contribute to feelings of pleasure. Ecstasy increases the amount of serotonin that is released into the brain, and long-term damage can be done to the brain as a result. Despite being chemically related to methamphetamine, Ecstasy is considered a Schedule I drug.

BARBITURATES

Barbiturates are synthetic *depressants* that account for a wide range of prescription and illegal drugs. A depressant affects the central nervous system of a user, causing feelings of peacefulness, sleepiness, and lethargy. Barbiturates are most commonly prescribed as sedatives and for general anesthesia, such as phenobarbital and secobarbital. In general, barbiturates are considered safe when prescribed by a doctor, but long-term users and abusers can develop addictions and tolerances to the drugs. Barbiturates are classified into schedules based on how fast they affect the body, where fast-acting drugs like secobarbital are considered Schedule II, moderate-speed barbiturates fall into Schedule III, and slow-acting barbiturates, Schedule IV. Barbiturates are most dangerous when combined with other depressants, like alcohol, the combination of which can lower the involuntary muscle functions like heartbeat and breathing, and can cause death. Other drugs that have the same effects as barbiturates are methaqualones, which are drugs like Quaaludes. Quaaludes are also a type of drug that sedates the central nervous system, but may also have addictive properties.

BENZODIAZEPINES

Benzodiazepines were developed as alternatives to barbiturates due to the high risk of dependency that was seen with barbiturate abuse. Unfortunately, benzodiazepines, just like barbiturates, have a heightened effect when mixed with alcohol, and the combination of the two can lead to coma or death. Common prescription benzodiazepines are Xanax, Librium, and Valium, and are typically all Schedule IV drugs. The most well-known and most abused benzodiazepine is Rohypnol, which is sold on the street as "roofies" or "date-rape drug." Rohypnol easily dissolves in liquid, and victims have no idea that they are consuming a drug, nor do they recall any of the previous events after they wake up. Due to the widespread illegal use of Rohypnol, it might be moved from a Schedule IV drug to a Schedule I drug because it has no other medical value.

ALCOHOL

Alcohol is a depressant on the central nervous system, and while alcohol use is not illegal (nor does it fall into any schedule on the Controlled Substances Act), long-term abuse can lead to a physical addiction. Alcohol is found in many forms, but mainly is consumed as beer, wine, or liquor. Alcohol intoxication is probably the most widely studied form of drug usage, and there is quite a bit of scientific research into the effects, metabolites, and detection of alcohol in a person.

Typically, alcohol intoxication occurs in stages, and is always based on various factors such as body weight, tolerance, and metabolism. The first stage of alcohol intoxication is a sense of relaxation and calmness. The second stage is a feeling of euphoria and loss of fine motor skills. The third stage is a lack of coordination and speech, and general disorientation about space and time. If the person continues to consume alcohol after the third stage, a complete depression of the central nervous system can result in death.

HALLUCINOGENS

Hallucinogens are substances that cause the user to experience an altered state of consciousness, which can be in the form of hypersensitivity to external stimuli (sounds, touch, sights), but can also cause the user to form a dissociative state from reality, or experience delirium. Common hallucinogens can be broken into two groups:

- **Naturally occurring** Substances such as peyote, which is a cactus that contains the drug mescaline, and certain *Psilocybe* mushrooms, which contain the drug psilocybin.
- **Synthetic** Chemical substances such as LSD (d-lysergic acid diethylamide), PCP (phencyclidine), and Ketamine.

Common dissociative/delirium hallucinogens are almost all naturally occurring, and are found in plants like deadly nightshade, mandrake, and jimson weed. The only synthetic dissociative drug is nitrous oxide, commonly known as laughing gas. The only animals known to have hallucinogenic properties are certain species of toads, the skin of which secretes a substance that contains the hallucinogen bufotenine.

Without exception, hallucinogens are universally classified as Schedule I drugs. While there is no known long-term health risk associated with hallucinogen use, there is a high level of psychological dependency that can develop with long-term use. Common street names of some hallucinogens are acid or microdots (for LSD), angel dust (for PCP), shrooms or magic mushrooms (for *Psilocybe* mushrooms), and Kat, Special K, or Vitamin K (for Ketamine).

MARIJUANA

Marijuana is a plant that contains high levels of the drug delta-9-tetrahydrocannabinol, or THC for short. Marijuana is also known generally as *cannabis*, which is the type of plant, as well as another street name for the drug. Marijuana, as a drug, typically consists of the dried, flowering female tops of the plant. The female portion of the marijuana plant contains resin glands, which is where the THC is highly concentrated. *Hashish* is a form of marijuana in which these glands have been separated from the plant, pressed into bricks or balls, and dried. While there is THC present in other parts of the marijuana plant, such as the leaves and stems, it is in very low concentrations, and often novice purchasers are sold the easily recognizable marijuana leaves instead of the female buds.

Marijuana is a Schedule I drug, despite recent research showing beneficial medical effects of THC on glaucoma sufferers and cancer patients. While the plant will probably remain as a Schedule I drug, derivatives and synthetic THC are currently being investigated for medical use and, if approved, would probably be granted Schedule II or III drug status.

The common street names for marijuana are grass, weed, bud, skunk, ganja, chronic, doobie, spliff, hash, herb, hydro, indo, maryjane, pot, and reefer.

Marijuana is commonly smoked in pipes, or rolled into cigarettes ("joints") or cigars ("blunts"). There are an incredible number of variations of pipes, water pipes ("bongs"), vaporizers, and other smoking implements for marijuana use. Marijuana can also be eaten, either raw or cooked into food, which could be anything from brownies to butter.

The effects of marijuana can vary greatly from person to person and from marijuana plant to marijuana plant, depending on the potency of the THC. Some users report anesthetic effects from use, while others report full-blown hallucinations. Commonly, users experience differential perceptions of space and time. While there are notable cases of physical and psychological addiction to marijuana, dependency of the drug seems to affect different users in different ways, and not all chronic users of marijuana are addicted to the drug.

Forensic Drug Testing

One of the most important factors to consider when working in a drug lab is the sampling methods that are used to test questioned drugs. While it might be desirable for a lab to test every gram of drug evidence that comes into the lab, fiscal and time considerations make such an objective infeasible. Because of the large number of drug cases that a forensic lab might receive, a sampling method must be employed that tests a representative amount of the questioned evidence. This is usually not a problem when dealing with a small baggie of questioned powder or suspected marijuana, but may be an issue when faced with a large shipment of bulk drugs, like a brick of cocaine or a backpack or garbage bag filled with heroin balloons. When faced with a large supply of suspected drugs, a drug chemist must test a representative sample of the entire set. While there is no set percentage or amount, a representative sample of a large seizure of drugs would be sampled and tested by qualitative and quantitative analysis (if appropriate in a given jurisdiction). If necessary, the results can always be duplicated with another round of testing. Duplication of the tests ensures that the first round of samples was representative, and allows for a better average to be established in the end for all of the suspected drugs in the seizure.

PRESUMPTIVE TESTS

When presented with a sample of an unknown drug, a drug chemist first conducts a *presumptive test* to determine what type or class of drug the sample might be. Presumptive tests are usually chemical-based, color-change tests, using a limited variety of different reagents. Presumptive tests, by their definition, are not specific for one type of drug, and can produce false positives and false negatives in the presence of other substances. Since presumptive tests are used only to screen a sample for a possible reaction, further tests are necessary if a positive presumptive test occurs. Toxicologists also use presumptive tests to screen for the presence of drugs or their metabolites. The most common types of presumptive tests used today in a toxicology lab are known as *immunoassay* tests. Chemical reagents are manufactured with antibody sites that recognize classes of drugs. When the immunoassay reagent is mixed in a solution with a drug or metabolite, a chemical reaction occurs that can be detected by a variety of methods such as the detection of radioactive labeled drugs or visible spectroscopy.

MICROCRYSTAL TESTS

Microcrystal tests use a very minimal amount of sample, and are performed quickly and inexpensively. Microcrystal tests are performed on a glass microscope slide, using a variety of reagents that react with the drug and form unique, specific crystal structures. With the advent of instrumental analysis, such as gas chromatography-mass spectroscopy, however, the majority of forensic labs use microcrystal tests as presumptive tests, and confirm identification through instrumental analysis.

CONFIRMATORY TESTS

Once the unknown or suspected drug has been narrowed down using presumptive tests, it is necessary to confirm the presence of the substance by using a *confirmatory* test. Confirmatory tests are more specific than presumptive tests and microcrystal tests, and by definition should narrow the scope of the findings down to one type, class, or form of a drug.

INSTRUMENTAL TESTS

Unlike microcrystal tests, which give fast results and are inexpensive, instrumental tests take more time and are more costly to conduct. Laboratory instruments are expensive, sometimes in the hundred thousand dollar range. Instrumental analyses are the most accepted way to prove the identity of the drug, which in order to prosecute is essential information to provide to the jury.

There is a wide variety of instrumental tests available to toxicology and drug labs. Most labs opt for one or two major instruments that can quantify a wide array of drugs. Most instruments cannot analyze raw samples. With toxicology samples, the tissue, blood, or urine sample must be prepared by extraction, purification, and filtration before it can be run on an instrument. Unknown drugs need to be extracted and dissolved in solvents that can be injected or analyzed on the instruments, and unknown liquids need to be purified to remove any chemicals that can damage the sensitive instruments. Once a sample is prepared, it can be run on a variety of different instruments using different protocols, each producing unique results that will help identify and quantify the unknown sample. The following are different types of procedures and instruments that are available in forensic toxicology and drug labs.

Thin-Layer Chromatography

Thin-layer chromatography (TLC) is a simple separation technique that uses silica-coated plates (the *stationary phase*) and chemical solvents (the *mobile phase*) to separate the different compounds of a drug. TLC is performed by spotting some of the extracted unknown sample onto the bottom of a silica-coated plate, and then immersing the end of the plate in a solvent inside a sealed glass chamber (see Figure 9-1).

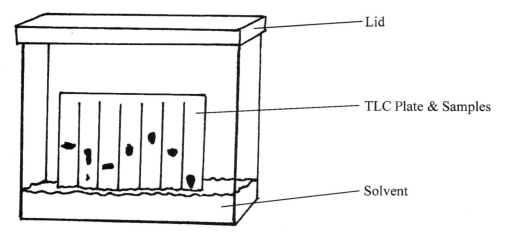

Figure 9-1   Thin-layer chromatography setup

As the solvent moves up the silica, the sample is carried away from the spot it was deposited and moves a relative distance on the plate based on the affinity of the chemical molecule to the stationary phase versus the mobile phase. TLC has a wide applicability, not just in drug testing, and is a valuable technique for separating components of a mixture (see the sidebar "Kitchen TLC").

Kitchen TLC

TLC can be mimicked easily in any kitchen. Simply take three different-colored pens and draw dots 1 inch from the edge of a paper towel. Next, pour a thin layer of rubbing alcohol into the bottom of a shallow pan and, holding the towel straight up and down, dip the end of the towel closest to the dots into the alcohol, allowing it to soak into the towel. As the alcohol is absorbed into the towel, it will run across the pen marks, and will start separating the dyes of the inks as the solvent moves up the towel!

Not only does TLC have the ability to separate substances, but in the case of drug analysis, the separated substances are still available for presumptive and confirmatory tests. The TLC plate can also be sprayed with presumptive chemicals to produce color changes, and the separated spots can be scraped off the plate, isolated, and run on more sensitive instruments for further analysis.

Ultraviolet Spectroscopy

This instrumental technique, known as UV-Vis or UV-visible spectroscopy, is performed on an electronic instrument known as a UV spectrophotometer. A spectrophotometer, in general, is made up of five distinct parts (see Figure 9-2):

- **Light source**   In the case of the UV spectrophotometer, this is an ultraviolet bulb, which emits UV radiation in the 230–340 nanometer (nm) range. UV light is defined as the region in the *electromagnetic spectrum* that falls within the wavelength range of 100–400nm. Visible light can be found within the 400–700nm range, gamma rays and x-rays fall at the most narrow end of the spectrum, and infrared, micro, and radio waves each have progressively longer wavelengths at the opposite end of the spectrum.
- **Focusing lens**   The focusing lens directs the UV radiation onto the sample.
- **Sample**   As the UV radiation interacts with the sample, a certain amount of energy is absorbed by the sample.
- **Monochromator**   The UV light that passes through the sample is then dispersed into the different wavelengths on a device known as a *monochromator*. The monochromator is a device similar to a prism or a slotted grating that separates the wavelengths of UV radiation and spreads them out, much in the same way a prism breaks white light into the primary colors.
- **Detector**   The dispersed light finally interacts with the detector, which is a photosensitive device that measures the intensity of the UV radiation that hits it. When a specific wavelength of UV radiation is less intense, it means that the sample absorbed it and it did not pass through to the detector.

The way in which a chemical absorbs UV radiation is directly related to the structure of the chemical, and since the structure of a chemical is what makes a chemical

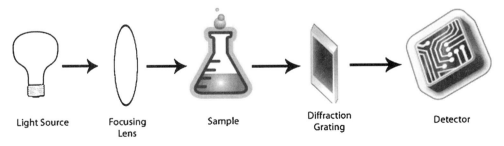

Light Source   Focusing   Sample   Diffraction   Detector
                  Lens   Grating

Figure 9-2   Basic spectrophotometer parts

unique, the UV spectrophotometer can identify specific chemical classes based on the UV absorption of an unknown.

Fourier-Transform Infrared Spectroscopy

This instrumental technique, known as FTIR spectroscopy, is performed on an electronic instrument known as a Fourier transform infrared spectrophotometer. The infrared region of the light spectrum falls between 700 and 1000nm, just outside of the visible region. The FTIR spectrophotometer uses infrared radiation to analyze mostly organic compounds. Based on the percentage transmittance of IR radiation, a drug's structure and composition can be determined because different frequencies are absorbed based on different molecular bonds inside the drug sample. Most FTIR machines are built with an interferometer, which allows for the rapid scanning of the entire IR range within a very short timeframe. The resulting data from an FTIR spectrophotometer is known as an *interferogram*, and when the mathematical calculation known as the Fourier transform is performed on this data, the resulting output shows the IR spectrum with the absorption bands from the sample measured in reciprocal centimeters (cm^{-1}), or *wavenumbers*.

FTIR spectrophotometer data displays the various chemical bonds that are present in a molecule, and can help uncover the various structures of an unknown sample. FTIR spectroscopy can be performed on solids, liquids, and even gases, but the preparation of the samples must be done carefully, and when dealing with solids, the sample must be completely dry, because the simple chemical bonds in a water molecule can mask the true sample data. Normally, when dealing with solid samples, the sample is pulverized into powder and mixed with potassium bromide salt. Potassium bromide is "transparent" in the infrared spectrum—it does not have any IR signature. The mixture of the sample and potassium bromide is then compressed under high pressure to form a sample disk. This disk is then placed in the FTIR spectrophotometer and scanned, and the spectra that are produced clearly show the sample data without interference from the potassium bromide.

Gas Chromatography-Mass Spectroscopy

The instrument that is used to perform this technique is actually two instruments linked together to form one extremely useful and powerful machine. Gas chromatography (GC) is another separation technique that uses a long narrow tube known as a *column* or *capillary*. At one end of the capillary is the injector port, and at the other end is the detector. The capillary itself is filled with an inert, porous substance (commonly silica), which is known as the *stationary phase*. When a sample is injected into the injector port, it is immediately vaporized and mixed with an inert gas, like helium, hydrogen, or nitrogen, and the gas mixture is then carried into the capillary. The inert carrier gas is known as the *mobile phase*. As the carrier gas pushes the sample along, the chemical components of the sample begin to separate based on the size of the vaporized compounds. Since the silica inside the capillary is porous, the smaller compounds quickly zip through and around the silica beads, and the larger molecules move slower. By the time the sample completely moves through the capillary, the small samples are the first components to elute off of the column and interact with the detector, followed by the medium-sized compounds, and finally the largest components (see Figure 9-3).

The detector at the end of the column "sees" the size of the relative molecule exiting the capillary, and graphs the intensity of the sample amount against the time (the time begins when the sample is injected into the first port). This is known as the *retention time* of a chemical or compound, and different chemicals and compounds all have different retention times.

After the different components of the sample are separated by the GC run based on their size, the components can be redirected into a second machine to uncover even more data. This machine is known as a *mass spectrometer* (mass-spec or MS for short). The MS instrument pulls in the separated components from the GC run and fragments the samples into *ions*. The ions are charged fragments of the original sample and can be detected with an amazingly high degree of sensitivity. The resulting data is a graph of the ions with the *mass to charge ratio* of the sample plotted against time. The MS data can then be compared against computer databases of known compounds to find a match.

Because the MS is capable of analyzing the mass of the sample, and the GC analyzes the separate components of the sample, the GC-MS combination is capable of absolute identification and possibly even individualization of a sample. For example, an unknown powder might presumptively test positive for cocaine using a

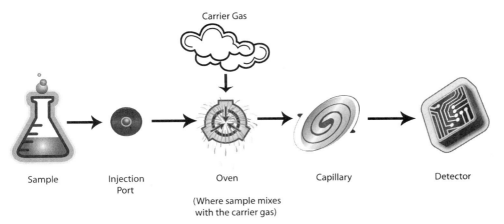

Carrier Gas

Sample   Injection   Oven   Capillary   Detector
                   Port

(Where sample mixes
with the carrier gas)

Figure 9-3   Basic gas chromatography process

color-change test. When this sample is injected into the GC-MS, the GC sample not only proves the presence of cocaine, but also can identify the other chemicals or compounds that were used to cut the cocaine before it was sold. When the MS interacts with the sample, it can not only again prove the presence of cocaine, but also identify the individual characteristics of the cocaine, along with the other adulterants in the mixture, and match this data to online data or reference data in a case to uniquely "individualize" the drug mixture. Normally, this type of "individualization" is not a true individualization because it only matches a known to an unknown based on several factors, but it cannot truly individualize the sample apart from all other samples. For this reason, GC-MS produces what is sometimes referred to as a *signature analysis*, which results in a unique GC-MS signature for that one sample that can be used to match against other samples or online libraries. Even though this is not a true "individualization" in the forensic sense, GC-MS does produce excellent identifications as well as unique signatures of substances. For this reason, GC-MS is known as the gold-standard instrument for forensic identification.

High-Performance Liquid Chromatography

The instrument that is used to perform this technique is much like the GC instrument described in the previous section, except that the sample is carried on a mobile liquid phase instead of gas. High-performance liquid chromatography (HPLC) tends to have a higher sensitivity and is faster than gas chromatography, but the sample preparation takes longer and there is a higher risk of damaging an HPLC instrument when injecting an unknown sample that might clog the column. Therefore, HPLC is often used to confirm GC results so that the proper liquids and conditions can be determined before the instrument is used.

Chemical Structure of Drugs

Certain drugs and chemicals have a right-handed and left-handed form, which means that although each chemical is exactly the same as far as the number and type of atoms are concerned, the left-handed form is a mirror image of the right-handed form of the chemical. These two forms of a chemical are known as *enantiomers* (see Figure 9-4).

This might not seem important, but it is! The human body reacts to certain left-handed drugs differently than the right-handed enantiomer of the same drug. For example, cocaine is a chemical that has an enantiomer. The left-handed form of cocaine produces the euphoric high, while the mirror-image enantiomer (the right-handed form) of the drug has a greatly diminished effect on the human body.

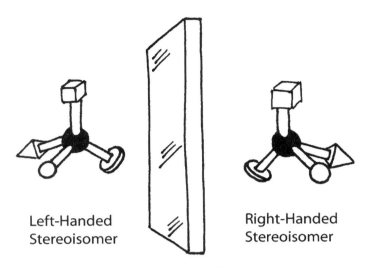

Left-Handed   Right-Handed
Stereoisomer   Stereoisomer

Figure 9-4   An example of an enantiomer

Unfortunately, the right-handed enantiomer of cocaine still has the same chemical structure and signature as the "real" form of cocaine, and will test positive in all presumptive and confirmatory tests in the laboratory.

There are several different instruments and methods in drug analysis that can uncover the actual structure of a chemical molecule. Sometimes, the two enantiomers of a drug can be seen at two different melting points when a mixture of enantiomers is tested for purity. One of the common ways to prove the presence of one enantiomer or a mixture of enantiomers is by examining the crystals using a polarized light microscope (PLM). Sometimes, the enantiomers can be isolated on special separation columns in a GC or HPLC. From an instrumental side, nuclear magnetic resonance (NMR) is a complex instrumental technique that is sensitive enough to determine the three-dimensional structure of almost any chemical compound. NMR uses strong magnetic fields to align atoms of a molecule in a certain way, and then forces them to rotate, vibrate, and spin by varying the strength of the magnetic fields. A *polarimeter*, on the other hand, is a simple, handheld instrument that is capable of determining the right-handedness or left-handedness of a chemical compound. The polarimeter is used to determine which enantiomer of a chemical is present, because one enantiomer of a chemical will rotate polarized light in one direction, while the mirror enantiomer of the same chemical will rotate polarized light in the opposite direction. Therefore, the only way to tell the difference between the left-handed form of cocaine and the inactive right-handed form is by using NMR or a polarimeter. If the NMR or polarimeter proves that the more potent left-handed form of the cocaine is present, then the case can proceed, because possession of cocaine is illegal. But, if the NMR or polarimeter proves that the inactive right-handed form of the drug is all that is present, then the case might be dropped based on the laws in a particular jurisdiction, because it may not be illegal to possess the "nonworking" form of cocaine! It is important to keep in mind, however, that this varies from state to state and jurisdiction to jurisdiction.

Poisons

Another important aspect of forensic toxicology is the detection and identification of poisons. Many of the same presumptive, confirmatory, and instrumental testing techniques that are used in drug analysis are also used to uncover the presence of poisons in suspected cases. For example, there are a variety of color tests and micro-crystal tests that can point to the presence of lead and cyanide in samples. Many of the same instrumental techniques are also used to identify and quantitate poisons. Poisons, by definition, are substances that cause injury or death to a living organism by either *acute*, *chronic*, or *latent* exposure. Acute exposure is defined as a one-time or short-time exposure to a harmful substance. Chronic exposure is simply exposure to the substance over a longer period of time, and usually there is less of the substance during each exposure but the effects over a long exposure time are cumulative. Latent exposure is caused by the toxic material entering the body but lying dormant in the tissues. The effects of this toxic chemical will not be present until another chemical or factor at a later time causes it to activate.

Another important factor in poisons is the dosage. At a very low level, minute amounts of arsenic will have little effect on a person, but a large concentration of the drug taken all at once can be fatal. The same is true with aspirin, alcohol, and other substances. As the famous medieval alchemist Paracelsus so keenly put it, "All substances are poisons; there is none which is not a poison. The right dose differentiates a poison and a remedy."

The final factor when talking about exposure to poison is the route of entry into the body. The routes of exposure apply not only to poisons but to all toxic substances: *inhalation* (breathing in fumes), *ingestion* (eating or swallowing the poisoned substance), *absorption* (through the skin or other mucus membranes), or *injection* directly into the bloodstream.

Many poisons that are discovered in cases of foul play are synthetic chemicals, such as drain cleaner and insecticides that have harmful effects when introduced into the human body. However, certain classes of poisons are naturally occurring and can be found in plants, animals, and the environment. The following sections describe various classes of poisons, both natural and synthetic.

PLANT TOXINS

Plant poisons are usually referred to as *toxins* and are produced naturally by plants in order to survive in whatever environment the plant grows. For example, strychnine is a toxin that is derived from the *Strychnos nux vomica* plant. The plant naturally produces strychnine to ward off animals from eating the plant. Strychnine is commonly used as rodent poison, and a very small amount can easily kill a human being.

Another not-so-common plant toxin is *curare*, which is isolated from plants in the South American rainforests and is used by indigenous tribes as the poison on the tips of arrows and blow darts. Curare causes complete paralysis when introduced into the bloodstream, including paralysis of the diaphragm, leading to suffocation and death.

Although it is technically not classified as a poison, *digoxin* is another substance that is extracted from plants. Digoxin is isolated from the digitalis plant, and it is mainly used as a drug for the treatment of heart conditions. Because the chemical nature of digoxin affects the heart and cardiac system of humans, an overdose of digoxin can cause a massive heart attack or other fatal coronary event. Remember, as previously mentioned, anything is poisonous in high enough doses. Digoxin was used in some key poisoning cases recently, including a nurse who was using it to "mercy kill" hospital patients in the late 1990s to early 2000s (see the sidebar "Charles Cullen").

Charles Cullen

Charles Cullen was a male nurse who worked in several hospitals in New Jersey and Pennsylvania, and was caught using digoxin overdoses to kill as many as 40 people. After the patients died from apparent heart attacks, the toxicology reports showed high levels of digoxin in their blood, which was never prescribed to many of the patients. Armed with this information, investigators linked several cases together with similar results and apprehended Charles Cullen in 2003 after he was identified accessing records and ordering medication for patients to whom he was not assigned. When presented with the toxicology evidence that linked him to several of the murders, Charles Cullen admitted to as many as 40 such murders that he had carried out at over ten hospitals since 1987.

ANIMAL TOXINS AND POISONS

Animals also produce poisons and toxins, most notably of which are snakes and spiders that bite and inject *venom* into their prey, causing death. These same poisons are also toxic to humans, although the most poisonous of spiders is harmless to humans. The common spider known as the daddy-long-legs has the most toxic venom out of all the spider species, but its fangs are not long enough to pierce the skin of a human being, and therefore their venom has no effect. However, there are spiders such as the black widow and the brown recluse whose bites can be fatal to humans. Snakes, such as the venomous pit vipers, rattlesnakes, and coral snakes, also all have venom that is fatal to humans.

ENVIRONMENTAL TOXINS AND POISONS

Environmental toxins and poisons can be found in a variety of forms, and most often people associate environmental toxins with chemical spills and improper waste management. While it is true that such "environmental" problems can lead to poisoning cases, there are also naturally occurring poisons and toxins in the environment that can also be fatal to humans.

The most common of all of these poisons is *carbon monoxide*, a colorless, odorless gas that is produced when organic compounds decompose or burn up. Carbon monoxide is fatal to humans in very minor quantities, because it displaces the oxygen in a room, leading to apnea, asphyxia, and death. A person being poisoned with carbon monoxide will feel tired and lose consciousness before realizing that there is anything wrong.

Another important naturally occurring poison is *cyanide*. Cyanide is an extremely powerful and quick-acting poison, and is found not only naturally throughout the environment, but also in industrial waste, dyes, and even raw foods. Cyanide kills by blocking the *electron transport* system in cellular systems. Because electron transport is responsible for the input of oxygen and nutrients and the output of carbon dioxide and waste, the organism that is poisoned with cyanide simply seizes up and dies due to asphyxiation.

HEAVY METALS

The final set of poisons to consider from a forensic standpoint is the naturally occurring heavy metals such as lead, thallium, arsenic, and mercury. Lead poisoning is seen less and less these days thanks to lead-free gasoline and paints, but lead can still contaminate groundwater systems, and chronic exposure can lead to brain damage, organ failure, and death. In ancient Rome, the first aqueduct system used lead pipes to bring water into homes. Since the Latin word for lead is *plumb*, this system became known as "plumbing." In the past, lead was also used in theatrical makeup, because lead oxide is a brilliant white, but direct absorption of the lead into the skin caused several actors to go insane, or become "plumb crazy."

Long-term exposure to mercury is also seen very infrequently these days, but can lead to serious brain damage. In the past, mercury was used in *haberdashery*, or the art of hat-making. Mercury was used to smooth out velvet and felt before being applied to a hat. The hat-makers' long-term exposure to mercury vapors left many of them brain damaged and insane, which developed into the cliché "mad as a hatter."

THE POISONS OF LETHAL INJECTION

Certain U.S. states carry the death penalty as punishment for particularly heinous or violent crimes. There is typically one clear-cut method that a state employs to legally terminate someone's life. Some states use the gas chamber or electric chair, but due to recent pressures from civil rights groups, many states have switched to a lethal injection. The process of lethal injection involves three chemicals in extremely high doses that are injected intravenously into the bloodstream of the condemned, one immediately after the other. The first chemical is thiopental sodium, which is a fast-acting barbiturate that causes the condemned to lose consciousness. Following the thiopental sodium, an overdose of pancuronium bromide is administered. Pancuronium bromide causes paralysis by relaxing all of the muscles in the body, including the diaphragm, and therefore breathing stops. The third chemical administered is potassium chloride, which causes an ionic imbalance in the heart tissue and induces cardiac arrest. With the breathing and heartbeat stopped, the unconscious condemned succumbs to death by asphyxiation within minutes.

Quiz

Refer to the text in this chapter if necessary. Answers are located in the back of the book.

1. Microcrystal tests:
   (a) Are somewhat specific
   (b) Require little time for the reactions to occur
   (c) Require a microscope
   (d) All of the above

2. UV spectrophotometers operate in the wavelength region:
   (a) Between 700 and 1000nm
   (b) Less than 100nm
   (c) Between 400 and 700nm
   (d) Between 100 and 400nm

3. The most common cutting agent in heroin is:

   (a) Rock salt

   (b) Gold-Bond's Medicated Powder

   (c) Oxy Clean

   (d) Quinine

4. Which of the following is not used in thin-layer chromatography?

   (a) Solvent chamber

   (b) Mobile phase

   (c) Stationary phase

   (d) Capillary column

5. Which of the following is the best technique for the identification of the chemical structure of an unknown drug?

   (a) GC

   (b) NMR

   (c) PCR

   (d) UV

6. HPLC uses which type of mobile phase?

   (a) Gas

   (b) Silica

   (c) Plates

   (d) Liquid

7. Which of the following is not a spectroscopic technique?

   (a) HPLC

   (b) UV

   (c) MS

   (d) IR

8. Peyote is harvested from:

   (a) Cacti

   (b) Mushrooms

   (c) Flowering bushes

   (d) Poppies

9. IR spectrophotometers operate in the wavelength region:

   (a) Between 750 and 1000nm

   (b) Less than 100nm

   (c) Between 400 and 700nm

   (d) Between 100 and 300nm

10. Which of the following is not an opioid?

   (a) Vicodin

   (b) OxyContin

   (c) Codeine

   (d) Valium

CHAPTER 10

Arson and Explosives

There are times during a forensic investigation that a forensic scientist might come across the scene of a fire or explosion. These types of scenes, by their very nature, are unique and must be processed in a particular manner, which requires specialized knowledge on the part of the forensic scientist. A crime scene involving a fire could be examined from the point of view that the fire itself was a crime. However, a fire could be used to cover up or destroy evidence of another crime, and care must always be taken when examining a crime scene where a fire was involved. Likewise, the use of explosives can cause considerable damage to a scene, and, like a fire scene, the explosion may be the main crime or may have been used to conceal another crime at the same time. Regardless, it is critical that a forensic scientist with the proper specialized training examine such scenes.

Arson Investigation

Locard's Exchange Principle states that every contact leaves a trace. As much as this is the foundation of forensic analysis, it has also become a well-known fact to the criminal element that perpetrates these crimes. Armed with the knowledge that their interaction with a crime scene leaves telltale clues that can lead to their down fall, criminals sometimes resort to drastic measures to reduce the presence of this trace evidence. Some criminals take precautions, such as donning gloves and wearing a hat. Some criminals attempt to clean up the crime scene after the fact; mopping up blood, picking up spent shell casings, or wiping their fingerprints off a gun. However, a select few criminals attempt to sterilize their crime with one of the oldest and most effective tools known to man, namely, fire. The act of purposefully and maliciously setting fire to an object or structure is known as *arson*. Arson may be done to cover up a previous crime, or it may be the main crime in and of itself. Arson can result in the destruction of a building or vehicle, and this might be done for thrills, insurance fraud, or even as a terrorist act. Regardless of the motivation behind the arson, it is still a crime, and one that requires forensic analysis.

While arson might cover up or destroy the trace evidence of one crime, Locard's Exchange Principle still applies to the act of arson. There will be traces of the arsonist and the arson method at the fire scene; a good forensic scientist knows where to look for these clues. A forensic scientist who specializes in arson investigation is known as a *fire investigator* or *arson investigator*. Sometimes, the arson investigator is not part of a police department, but instead is a member of a fire department. After every major fire, a *fire marshal* is responsible for investigating the burnt remains of a building, structure, or vehicle. Based on the clues, the fire marshal determines whether the fire was natural, accidental, or arson. What a fire marshal is looking for at the scene of a fire are patterns, traces, and evidence indicative of arson, using the common forensic methods. As a result, the fire marshal and arson investigator are doing the same job, the only difference being that one works for a fire department and one works for a crime lab or police department.

THE FIRE TETRAHEDRON

To better understand the job of an arson investigator, it is first necessary to understand the nature of fire. *Fire* is defined as a *rapid exothermic oxidation reaction*. What that really means is that fire is a quick, heat-producing process that requires oxygen. But, oxygen is not the only ingredient in fire. In fact, for a fire to occur, there must be four factors satisfied. These four factors make up what is known as the *fire tetrahedron* (see Figure 10-1).

Figure 10-1 The fire tetrahedron

The factors that must be present for a fire to occur are oxygen, a fuel source, heat, and a sustained chemical reaction. (In the past, this was referred to as the *fire triangle* and did not include a chemical reaction. However, recent changes in the understanding of fire science required the addition of *chemical reaction* to the list, to deal with chemical fires that need only a miniscule fuel source to ignite, followed by a sustained chemical reaction.)

The main idea behind the fire tetrahedron is that if one of the factors is removed or eliminated, then the fire goes out. Therefore, the fire tetrahedron is used not only in fire science, but in fire fighting. When firefighters spray water onto a fire, it cools the fuel source, removing heat from the equation, and the fire goes out. When a fire extinguisher is used, it emits either a layer of fire-retardant foam or carbon dioxide, both of which displace oxygen away from the fuel source and thereby extinguish the fire. For chemical fires, such as burning magnesium, if the chemical reaction can be balanced or reversed, then the fire will go out as well.

There are many materials, both natural and synthetic, that are easily flammable. Most materials that burn do so because they are *organic*; that is, they are made up of carbon. Coal, which is essentially nothing but carbon, has been used as a fuel source for hundreds of years for this very reason. Inorganic substances, on the other hand, are usually not flammable, but certain elemental forms of some chemicals, such as magnesium and sodium, can "ignite" when exposed to water or other oxidizers. Inorganic substances contain no carbon, and therefore do not burn. Asbestos, an unfortunate cancer-causing agent, is made up of inorganic mineral fibers. These fibers are fireproof, and this is the reason why asbestos was used so extensively as a fireproofing insulation until it was banned in the mid-1980s. Asbestos has been largely replaced with gypsum, fiberglass, and various other fireproof synthetic materials, but they all share the main principle of remaining inorganic.

ACCELERANTS

Fire can start spontaneously or from a small single source. Thanks to a wonderfully effective U.S. Forest Service advertising campaign using a talking bear, many people know that "only you can prevent forest fires." A single stray cigarette was blamed for several accidental forest fires in the 1990s, which is a clear example that Smokey the Bear and his friends knew what they were talking about. However, fires can also be "helped" along. Anything that is used to speed up the creation of a fire or to make the fire burn hotter than normal is known as an *accelerant*.

Accelerants can range from a variety of petroleum-based liquid chemicals (such as gasoline or butane) to flammable gases (like natural gas), and even some solvents (such as turpentine or isopropanol). There are even reports of bags of potato chips that were used as accelerants in automotive arson fires. The accelerant works by increasing the fuel available to the fire and thereby increasing the heat at which the fire burns. Sometimes, accelerants are present at accidental fires, such as a stray cigarette at a gas station or a grease fire in a kitchen. More commonly, however, accelerants are used in arson fires to quickly spread the fire and maximize the damage. It should be noted that in common parlance, the term "ignitable liquids" has superseded the term "accelerants," but for the purpose of our discussion about arson investigation, we use the terms interchangeably.

RETARDANTS

To counter the effects of an accelerant, or to slow down or stop a fire, a *retardant* is used. A retardant is anything that limits or cuts off one of the four factors in the fire tetrahedron and controls the fire. Fire can be controlled in various ways. A cigarette lighter or a Bunsen burner is a good example of a controlled fire. In a cigarette lighter, the fuel source is controlled by the person holding down the trigger, allowing the butane gas out of the spigot. If the trigger is released, the gas stops and the flame will extinguish. A Bunsen burner has an adjustable flue, to increase or decrease the amount of oxygen that is mixed with the natural gas to increase or decrease the heat and size of the flame. These, however, are physical retardants. Most commonly, retardants are chemical in nature and work by limiting the oxygen to a fire or cooling a fire by limiting the heat. While retardants are mainly used to fight fires and help put them out, sometimes retardants are also used illicitly to delay a fire from occurring immediately, which allows the arsonist time to leave the building or structure, and as a result there might be enough time to establish a credible alibi when the fire actually breaks out. In this manner, a retardant is applied to a small fire or used to control a fire, such that when the retardant is used up or dissipates, the fire grows bigger or reacts with an accelerant to rapidly burn.

IGNITION SOURCES AND HEAT

Even with the use of accelerants and retardants, fires do not start by themselves, even in nature. A naturally occurring forest fire still needs a spark or a lightning strike to begin the ignition of the wood or dry brush. The same applies to arson. To light a fire, an *ignition source* is necessary. This can be as simple as a match or as complex as a timed electronic fuse.

The type of ignition source depends on the overall flammability of the fuel source. A highly flammable gas like butane needs only a small spark to ignite. A log of wood requires a much longer and constant ignition source, such as a sustained flame from a blowtorch (which, unfortunately, is actually done by some boy-scout dropouts to start campfires). Whatever the source of ignition, it provides the first factor, *heat*, which results in the sustained chemical reaction that is fire. Therefore, the ignition source must be applied directly to the oxygenated fuel source in some way. In other words, if you want to light a pile of newspaper on fire, you hold a match to it. If you are an arsonist and want to burn down a building, you need to set fire to the accelerant or the main structure in some way.

Arson Investigation and Reconstruction

The forensic investigation of an arson scene proceeds much the same way as any other scene is processed—photographs are taken, notes are made, and evidence is searched for and collected. In an arson investigation the evidence the arson investigator is looking for is twofold. First, an investigator at any fire needs to locate the *origin* of the fire. This indicates where the fire started, and could either help rule out arson or provide strong evidence for it. Interestingly enough, fire has a way of leaving a trail that points back to the origin of the fire. Just as heat tends to rise, so does fire, and it spreads out along the way. If fire is therefore moving up a wall, it spreads out and forms a V-shaped burn pattern (see Figure 10-2).

The bottom point of this V pattern literally points directly to the origin of the fire. If this happens to be an electrical socket, a quick examination might reveal faulty wiring or a short that caused the blaze, and the fire can be ruled accidental if no other indicators of foul play are uncovered. However, if the V-shaped burn mark indicates an origin with no clear ignition source, an arson investigator might have

"V" Shaped Burn Pattern

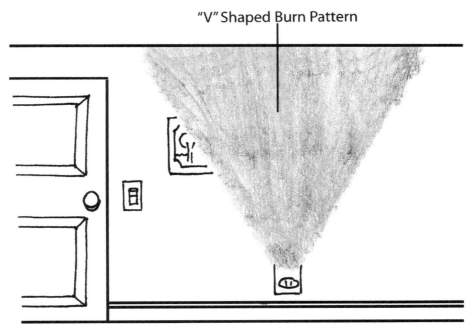

Figure 10-2   An example of a V-shaped burn

to consider that arson could be the cause of the blaze, but how does one prove it? This brings us to the second item of fire investigation.

If arson has the ability to cover up previous crimes, can it not also cover the trail of an arsonist? Incredibly, no. Accelerants and retardants tend to be liquid or powders, and therefore settle to the lowest point in a space according to gravity. As the heat of fire rises, any leftover accelerant or retardant used to start the fire might be spared from ignition. In most cases of arson, this means that the accelerant or retardant might soak into carpet, rugs, furniture, or flooring. Because of the *volatility* of accelerants, they also evaporate quicker and these fumes can then soak into walls, wood, and other items in a room or space. For an arson investigator, this means that the evidence of arson must be found quickly, but if it is there, it is not visible to the naked eye, and it is rapidly evaporating. This problem is compounded by the fact that the scene might be structurally unsound from the fire, or that the process of fighting the fire added more chemicals or retardants to the scene.

DETECTION AND COLLECTION OF ARSON EVIDENCE

If arson is suspected at a fire, an arson investigator can use trained dogs or special instruments called "sniffers" to try to locate any accelerant or other chemical in the remnants of the fire. These methods, the dogs or the sniffers, are simply a presumptive screening for the presence of accelerants, and should be used only to locate them at a crime scene.

Once an area with a suspected accelerant is identified, a sample of that area must be taken and sealed in an airtight, clean metal canister. The ideal container for this purpose is a new, unused paint can. Either the sample from the scene is placed in the container by itself or a piece of *activated charcoal* is added along with the sample (more on that later). Whatever the material that is sampled, a similar piece of the same material located away from the site of suspected accelerants should be sampled as a control. For example, if the corner of a carpet in a room is suspected of containing accelerants, the opposite corner of the same carpet needs to be screened and, if negative, sampled as a control. The control sample will help account for the numerous chemicals that are already present in the carpet sample, including those which might have been "released" by the heat of the fire. Also, the control must be treated the same way as the suspected sample. Both samples should be placed in separate metal canisters, sealed (if activated charcoal is added to one, the other must have it as well), and transported immediately to the lab.

So what's the deal with activated charcoal? Why would an arson investigator add it to the canister? The reason is, simply, that it can provide an alternate means of isolating any volatile accelerants that might be in the sample. Activated charcoal is highly adsorptive and will trap any volatile liquids or gases that it comes into contact with. The activated charcoal that is added at the scene therefore needs to be sterile and come from a clean source, like a lab or company that specializes in the production of clean activated charcoal. If activated charcoal is used, it provides one way of sampling the volatile liquids or gases that may be soaked into the sample in question. If it is not used, then these volatile gases will be trapped in the environment inside of the metal can, and must be sampled in another way.

LABORATORY ANALYSIS OF ARSON EVIDENCE

Once the canisters are back at the lab, the contents must be sampled to screen for the presence of accelerants, retardants, or other chemicals. The most common approach to this method is the use of the gas chromatograph-mass spectrometer (GC-MS). The gas chromatograph is extremely good at separating the different components of a mixture based on size. Since most accelerant agents are based on some sort of hydrocarbon chain or chains, the identification of these structures can lead to proof that an accelerant was used, as well as identify the type of accelerant. The mass spectrometer can also prove the presence of an accelerant, but, more importantly it can analyze the minor components that make the accelerant unique, such as impurities or other compounds, and match this data to a known sample, library, or database. Once the individualizing features of the accelerant are known, it is possible to identify the brand of accelerant used or even where it was originally purchased.

Accelerants are classified into groups for identification purposes. There are six main classes of accelerants, listed next. Since most accelerants and ignitable liquids are the products of petroleum refining, the classifications are grouped according to the "weight" of the petroleum chemicals that make up the accelerant.

- **Class 1**   These accelerants are known as *light petroleum distillates*, and include butane and other types of lighter fluids.
- **Class 2**   This class is reserved for gasoline only, but includes subcategories for all the different octane ratings of different gasoline.
- **Class 3**   These compounds are grouped as *medium petroleum distillates*, which mainly include charcoal lighter fluids and mineral spirits.
- **Class 4**   This class is reserved for kerosene and other fuel oils, such as home heating oil.
- **Class 5**   These compounds are grouped together as *heavy petroleum distillates*, and include such heavy flammable liquids as diesel fuel.
- **Class 0**   This is a catch-all classification for other ignitable liquids or accelerants that do not fall into one of the other five categories. Compounds that fall into this miscellaneous category are chemicals such as cleaning solvents, candle oils, and lacquer thinners, as well as the liquids used in aerosol can sprays.

Accelerant Sampling in the Laboratory

To identify these accelerants and figure out the class, however, there is one hurdle that must be overcome. How does one get a sample of volatile gas out of a sealed metal container and into the sample port on a GC-MS? One of the solutions to this problem is to use activated charcoal. If the metal sample canister remains sealed and is heated, the suspected accelerants in the sample will volatilize quicker. As a result, more of the accelerant will be adsorbed by the activated charcoal. The charcoal can then be removed from the canister, and, using a series of solvents, the accelerant chemicals can be extracted from the charcoal matrix using a solvent. Once in solution, this accelerant-rich solvent mixture is injected into a GC-MS for analysis, identification, and, individualization. This type of extraction is known as a *passive concentration* of the sample, because it relies on the charcoal to randomly adsorb whatever it comes into contact with inside the canister.

To ensure a more efficient sampling of the suspected accelerants, activated charcoal can again be used, but it is not directly inserted into the sample canister. Instead, the canister is heated and punctured with two gas lines. One line pumps in a gas, such as nitrogen, while the other line sucks the volatilized accelerants out of the can as they are displaced by the nitrogen. As these sample gases are removed they are forced

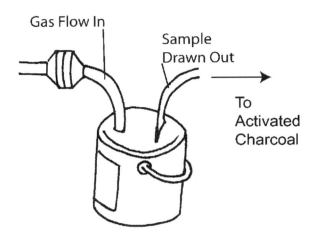

Gas Flow In

Sample
Drawn Out

To
Activated
Charcoal

Figure 10-3   Dynamic concentration for sampling arson evidence

through an activated charcoal filter (see Figure 10-3). Once the suspected accelerant gases are trapped in the charcoal, they can be extracted into a solvent, which is then injected on the GC-MS. This type of sampling is known as *dynamic concentration*, due to the nature of forcing the entire sample out of the canister and into the charcoal. This method tends to be highly effective in the analysis of arson accelerants.

If, on the other hand, activated charcoal is not used to adsorb the suspected liquids and gases, then these gases will be encased inside the canister instead. To extract these gases, the metal sample canister again will be heated, furthering the release of even more of the volatile gases into the interior of the canister. At this point, two other methods are available for sampling suspected accelerants. First, the container can be punctured with a syringe, allowing the interior air to be sampled directly and injected into the GC-MS (see Figure 10-4). This is known as *simple*

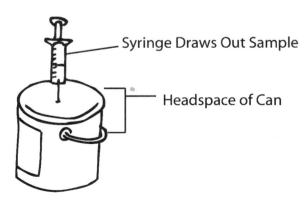

Syringe Draws Out Sample

Headspace of Can

Figure 10-4   Simple headspace for sampling arson evidence

headspace sampling, because the air above and surrounding the sample, (the "headspace") is being directly sampled. This tends to be an easy and reliable method for sampling, but if the syringe needle samples the gases from only one layer of the can, heavier components of the sample at the bottom of the can might be missed.

Simple headspace sampling should only be attempted once all other sampling methods have proven unsuccessful. If simple headspace testing fails to identify any sample, then the evidence sample itself can be removed from the can and washed with solvents. These solvents will remove any suspected accelerant along with any other chemicals trapped in the sample, in the process rendering the sample unusable for further testing. The solvent wash can then be injected into the GC-MS to screen for the presence of any accelerants. This method is known as *direct solvent extraction*, and it is problematic for many reasons. First and foremost, fire debris is dirty, and a GC-MS is a sensitive instrument. The sample must be purified and filtered before the extracted solvent can be injected into the machine, and as a result some of the sample might be lost in the cleanup or filtering process. More importantly, the solvent removes all of the volatile chemicals from the sample, not just suspected accelerants. Most synthetic materials these days have remnants or trace levels of volatile chemicals present in them, hence that "new car smell" in a brand-new automobile, for example. What you really smell in a new car is the volatile compounds from the seat and floorboard fabric, the padding in the seats, and the plastic finishes or wood varnishes on the various parts of the interior of the vehicle. Therefore, when a direct solvent extraction is performed on a synthetic substance, all of the chemicals that make up that substance are removed and injected into the GC-MS, not just the accelerants.

THE INTERPRETATION OF ARSON ANALYSIS RESULTS

In the end, the sample data is meaningless unless compared against two other pieces of data. The control sample that was also collected from the scene must be sampled in the exact same way the original suspected sample was tested. Any background chemicals or native volatile mixtures can then be identified in the control and ignored in the sample. This is especially important if dealing with direct solvent-extracted samples, where the original chemicals must be accounted for by also extracting the control. Once the similar chemicals from the control are accounted for and removed, what remains is to identify and individualize the accelerant, if present.

As previously mentioned, the GC-MS data from an identified accelerant can be compared against a computerized library or online database to a given ignitable liquid. It is not possible to determine a given brand or a specific lot number. Interpretation problems are caused by a variety of issues. An ignitable liquid may be weathered; that is, some of the more volatile components may have evaporated from the unburned mixture, rendering the liquid sample found at the scene or in possession of the arsonist different from a standard sample. Furthermore, when materials such as plastics, wood products, carpets, etc. are burned, they go through a distillation-like process and chemicals that might be confused with ignitable fluids could be present. For these reasons, the interpretation of analytical results may be very difficult in some suspected arson cases.

In the end, a fire might only take seconds to start and within minutes cause significant damage. The reconstruction and analysis involved in an arson investigation might take days, weeks, or even months to uncover the presence of accelerants and match them to a known sample. However, if the samples are not collected soon after the fire, there might be little else to go on, and the crime may go unsolved. As is always the case in forensic science, but especially true in arson investigation, time is never on the side of the forensic analyst.

Explosives

The forensic investigation into explosives goes hand in hand with arson investigation. By definition, an *explosive* is any material that is chemically unstable and, if exposed to some type of shock, will undergo a rapid thermal expansion. An *explosion* is the result of the detonation of an explosive. An explosion is characterized by two aspects: the generation of heat, and rapid expansion that results in a "shockwave." Explosives are grouped into classes according to the speed of the shockwave. A *low explosive* is an explosive substance that results in an expansion traveling at less than 1000 meters per second (<1000 m/s). Common low explosives are substances such as gunpowder and the propellant used in fireworks and model rockets. These explosives are also very stable and require an energetic shock for the reaction to occur. This reaction is known as *deflagration*, which is the burning of the explosive rather than the expansion of it.

A *high explosive*, on the other hand, is classified as an explosive material that results in a shockwave traveling faster than 1000 meters per second (>1000 m/s). The reaction of a high explosive is known as *detonation*, which is characterized by the rapid thermal expansion of the material. High explosives are usually used to trigger low explosives, and as such high explosives are further broken down into three categories:

- **Primary high explosives** Explosive substances that are extremely unstable and highly susceptible to shock. Such explosives are dangerous in large quantities, and as such are usually kept only in small amounts. A common primary high explosive is the primer charge found in the primer cap on ammunition. The shock of being crushed with the firing pin causes the primer charge, which is usually lead azide or potassium perchlorate, to rapidly detonate. This explosion triggers the ignition and detonation of the low explosive gunpowder and, as a result, the gun fires.

- **Secondary high explosives**   Explosive substances that are less susceptible to shock than primary high explosives, and are therefore more stable. However, a quick, hard shock can easily detonate a secondary high explosive. Common secondary high explosives are substances such as RDX (also known as C4 or plastic explosive), dynamite (also known as TNT), and PETN (pentaerythritol tetranitrate). These secondary high explosives are commonly used as booster charges for larger explosive devices, such as landmines, artillery rounds, and detonators.
- **Tertiary high explosives**   Explosive substances that are extremely stable, to the point where they cannot be detonated without a detonator or booster charge. Despite not being sensitive to shock, these explosives still produce a thermal expansion of gases in excess of 1000 m/s, and in the case of the tertiary explosive ANFO, can greatly exceed this speed (see the sidebar "ANFO").

ANFO

This tertiary high explosive is named simply for its components, ammonium nitrate and fuel oil. This explosive is also known as a fertilizer bomb, due to the fact that ammonium nitrate is a common agricultural fertilizer, as well as a powerful oxidizer. When mixed with fuel oil, such as diesel fuel or kerosene, the resulting slurry is highly stable and yet incredibly destructive when detonated. While there are legitimate uses for ANFO in the mining and construction trades, its use is more widely known for the destruction of the Murrah Federal Building in Oklahoma City in 1995, and in the original World Trade Center bombing in 1993.

THE EXPLOSION CRIME SCENE

Many police departments in the United States do not have the capability to reconstruct bomb scenes. Typically, larger municipal agencies and state agencies have these capabilities, and federal agencies are available to assist as well. Major bombing incidents often become federal investigations and fall under the jurisdiction of the FBI or the Bureau of Alcohol, Tobacco, Firearms and Explosives (ATF). Regardless of the agency conducting the investigation and analysis of a bombing scene, investigators and forensic scientists try to reconstruct the event and obtain as much information to determine who was responsible for the bombing.

Bombings create major damage and destruction, which adds to the difficulty of sorting out what happened. With any bombing or explosion, there is a correlation between the amount of destruction and the size of the explosive device. The bigger the blast, the more work that must go into the reconstruction of not only the object that was destroyed, but also the events surrounding the bombing and even the makeup of the explosive device itself.

The reconstruction of the object that was destroyed is usually accomplished by finding the scattered pieces and reconstructing the object like a puzzle. When Pan-Am Flight 103 was bombed in midair over the Scottish town of Lockerbie on December 21, 1988, pieces of the airplane were scattered over a 2-kilometer area. The plane was brought down with a plastic-explosive bomb hidden inside of a Toshiba portable stereo, which was in the luggage compartment when detonation occurred. The reconstruction of the pieces of the wreckage took several months, but in the end, the Boeing 747's fuselage was rebuilt in an aircraft hanger. The examination of the rebuilt fuselage showed the investigators that a bomb indeed had detonated inside the luggage compartment, evidenced by a huge hole in one section of the rebuilt plane. Also, due to the careful collection and reconstruction of the other components inside the airplane, the investigators were able to show exactly what piece of luggage the bomb was in, and exactly where that suitcase was stored in the luggage compartment. In doing so, the investigators in the Lockerbie case uncovered the *center of the blast*, which is a critical step in any explosion reconstruction. This area was identified as the origin of the explosion of all the surrounding metal that was bent outward, as well as charred.

In any case, once the center of the blast is established, a measure of the damage of the explosion can be made. Explosive devices detonate and expand in 360 degrees unless acted upon by a stronger force. Therefore, any detonation that is unimpeded should cover a circular area completely surrounding the center of the blast. Sometimes, there might be two zones of destruction from an explosion. The innermost zone is the result of the *shrapnel* (see the sidebar "What Is Shrapnel?") that is cast off from the explosion, and the furthest zone is the destruction caused by the energetic *shockwave* from the blast. The distance from the center of the blast to the outer edge of the blast zone is known as the *blast radius*, and is a measurement of the explosive force of the bomb. However, the measurement of this zone does not tell an explosion investigator what the bomb was made of.

What Is Shrapnel?

Shrapnel is defined as any kind of projectile that is cast away from an explosion. The effectiveness of a bomb's shrapnel is known as the *brisance*, or "breaking ability," of the bomb. The shrapnel that is cast off from an explosion might be the outer case of the bomb (as in a hand grenade) or objects purposefully packed around the explosive material, such as the use by suicide bombers of screws, nails, and/or ball bearings surrounding explosives. The brisance of a bomb can be established when reconstructing an explosion scene by examining the recovered shrapnel. The quality of the size of the shrapnel can then serve as a clue as to whether the device was homemade or not.

The blast radius can be used to determine the amount of explosive used, but only if the type of explosive is known. For example, if a low explosive like gunpowder were packed into a sphere the size of a tennis ball and detonated, there would be a sizeable blast area, probably 10 to 30 feet wide depending on the quality of the bomb. However, if the same size tennis ball were packed with a high explosive, like plastic explosive, and detonated, the resulting blast radius might be 10 to 50 times larger. Same-size explosives, therefore, create different results based on the type of explosive used. To determine exactly what type of explosive was used, samples must be collected and tested for *explosive residue*.

EVIDENCE RECOVERY AT THE SCENE OF AN EXPLOSION

Explosive devices are never 100 percent efficient, meaning that not all of the explosive material in a bomb is detonated when the explosion occurs. As the explosive material itself expands, microscopic traces of the intact explosive are scattered over the same area that the blast occurred in, creating a "blanket" of residue of the unexploded charge over the blast area. Since explosives are chemical based, this residue can be readily collected, tested, and identified using standard forensic methods. The very nature of this testing is almost exactly the same as the tests and sampling methods used in arson investigation, which is why these two specializations are grouped together in forensic terms.

The first step in the recovery of explosive residue is the collection of the material from the explosion site. This can be done by collecting every fragment and swabbing it to collect possible residue, or by random sampling. Generally, explosive residue can be readily collected on cotton swabs or patches. Just as most explosives are stable under optimal conditions, so too are the residues dispersed from the explosion. Also, because the amount of unreacted material in explosive residue is so small, the effect of any further detonation would be minimal or unnoticeable. Unlike arson evidence, most explosive residues are not volatile and therefore should not evaporate. However, proper packaging is still necessary, and new metal canisters can again be used, as well as special plastic bags designed for the preservation of explosive residue evidence. These bags will not transfer any plastic or chemical signature back to the sample, and are easier to use and store than metal containers.

ANALYSIS OF EXPLOSIVES EVIDENCE

The presumptive testing of explosive residue can be done in many ways. There are simple color-change reagents that can be used to screen for the presence of certain explosives. Microscopically, there are several ways to identify explosive residue, from the use of microcrystal tests, to identification of optical properties using polarized light microscopy. Just as in arson investigation, there is also a variety of chemical sniffers and electronic detectors that can be employed at a scene to search for and screen for the presence of explosive residues. Finally, there are even portable instruments, such as a portable GC-MS, that can be set up at an explosion scene and used to quickly screen evidence for the presence of explosive residue. These "portable" systems are also commonly seen at airports, where Transportation Security Administration agents will swab the metal components of a piece of luggage with a small square of cloth and place it in such a machine. This machine, called an explosive trace detection system (ETD), is a rapid-scan mass spectrometer that is set to screen for eight to ten common explosive signatures. Each test on such a machine takes 10 seconds or less, so samples can be quickly swabbed and screened with a fairly high degree of accuracy, but this is only still a presumptive test. Even though very few substances can result in a false positive on such a screening test, all samples must be verified using standard instrumental techniques in the lab. Therefore, the identification of the explosive can be made with GC-MS or with high-performance liquid chromatography coupled with mass spectroscopy (HPLC/MS). Keep in mind also that the sensitivity of these instruments is high enough to detect the unique components of the explosives, and if this data is compared against a known standard or database library, individualization of the explosive is possible. However, this is again a "weak" individualization, much like the individualization of arson accelerants, and the results might only point to a specific brand, make, or lot of the explosive.

The results of the identification of the explosive can then be used, along with what is known about the blast radius, to reconstruct the amount of explosive that was used in the bomb. However, other factors must also be taken into consideration when examining the efficiency of the explosion. Sometimes multiple explosions happen in a pattern, designed to propagate each blast into the next, with the final shockwave having a total destructive effect of all of the individual explosions. As a result, a number of small explosions might result in *more* destruction than one large explosion. Also, a sensitizer can be added to an explosive compound to increase the effect of the blast. A *sensitizer* is any inert substance that, when dispersed by the explosion, creates voids in the air, allowing the shockwave of the blast to travel farther by not being slowed down by air pressure. In other words, the shockwave rides in the "wake" of the sensitizers and moves faster and farther as a result. This too can lead to more destruction at the scene of a bombing. However, this sensitizer, like the explosive residue from the bomb, will be evenly distributed over the blast radius, and if collected, it can be identified and "added" to the equation for the blast efficiency. In the end, the reconstruction of an explosion scene is done to prove that a bomb or other explosive device was used, and that the explosion was not "natural." However rare, natural explosions do occur, and can be something as simple as a meteorite strike (the results of which resemble a bomb crater, except for a total lack of explosive residue) to something as complex as a grain elevator explosion. Although there are better safeguards in place today, occasional grain elevator and silo explosions still happen every year. These are the result of highly flammable grain dust mixing with air in a large, confined space like a silo. The evenly dispersed dust in the air makes it susceptible to instant incineration (remember the fire tetrahedron—fuel + oxygen), except that the rapid thermal expansion inside the silo causes the silo to explode like a bombshell. Again, the reconstruction of this explosion, if natural, will uncover no trace of explosive residue, although it cannot rule out arson.

The end results of the investigation into an explosion can therefore piece together the events, methods, and devices that were used to create the explosion. An identification of a particular explosive or the individualization of a unique component of an explosive compound can point an investigator to the source of the explosive material, whether that is the local hardware store, a black-market military weapons stash, or a chemical company. Only by thorough analysis of all the data from the investigation, not just the explosive residue and blast radius, can a final conclusion be reached and, hopefully, a suspect or suspects identified and arrested.

The Reconstruction of the 1993 World Trade Center Bombing

The destruction of the World Trade Center towers on September 11, 2001, was arguably the most destructive and lethal terrorist strike ever committed on American soil, but it was not the first time the World Trade Center had been attacked. Roughly eight years earlier, on February 26, 1993, a car bomb was detonated in the lower-level parking garage at the World Trade Center, under the north tower. The explosion was caused by a urea-nitrate bomb, otherwise known as an ANFO (ammonium nitrate fuel oil) bomb. The bomb's gross weight was estimated to be around 1500 pounds, based on the size of the blast radius. The damage caused by this device resulted in a 20,000-square-foot hole that extended seven stories through the underground garage structures of the World Trade Center. The attack was designed to weaken the foundation under the south-eastern corner of the north tower, which in theory would have caused the north tower to collapse, falling into the south tower as it came down, destroying both towers in the process. Luckily, in 1993, this did not happen. The shockwave from the ANFO bomb, while succeeding in punching through the thin concrete layers from one level of the garage to the next, was not as effective at destroying the massive steel-reinforced concrete pillars that supported the building. What could have been a horrible tragedy was averted by poor planning on the part of the bombers. However, there was still extensive damage from the blast, which resulted in six deaths and thousands of injuries (see the sidebar "The 1993 WTC Memorial").

The 1993 WTC Memorial

The victims of the 1993 bombing were commemorated with a fountain and reflecting pool placed at the street level of the north tower. However, this memorial itself was later destroyed in the terrorist attacks on September 11, 2001, which ultimately destroyed both World Trade Center towers as well as the surrounding area. A single piece of this fountain, measuring only a couple of inches long and inscribed with a partial name, was the only piece of the 1993 memorial that was recovered following 9/11. This fragment was remounted in a second memorial on the south side of Ground Zero, in the viewing area where families of the 9/11 victims could come to pay their respects.

Due to the "failed" nature of the 1993 bombing, coupled with the fact that the remnants of the bombing were trapped within the parking garage, the forensic reconstruction of the bombing itself was carried out in a fairly straightforward manner. Between the results of the explosive residue analysis and the area of the destruction, the size of the ANFO bomb was determined to be approximately 1500 pounds. Also, corroborating security camera footage clearly showed the van that was used to carry the bomb not only entering the garage that day, but driving around lower Manhattan as well. By recovering key elements of the vehicle, as well as reconstructing the license plate numbers from the camera footage, a trail emerged that led investigators back to the terrorists that had rented the van, delivered the bomb to the World Trade Center garage, and set the timers on the detonators. The total investigation took over five years, but in the end, six men were convicted of the bombing. Between the trials in 1997 and 1998, each terrorist was found guilty, and was given a 240-year sentence for conspiring to destroy the World Trade Center.

Quiz

Refer to the text in this chapter if necessary. Answers are located in the back of the book.

1. Fire debris evidence should be packaged in:
    (a) Airtight plastic bags
    (b) Brown paper bags
    (c) Used paint cans
    (d) Steel cans

2. Which of the following techniques is best for the unique identification of a volatile compound?

   (a) FTIR

   (b) UV spectroscopy

   (c) PCR

   (d) GC-MS

3. What is the indicator of the origin of a fire at a crime scene?

   (a) Liquid on the floor

   (b) An empty gas can

   (c) V-shaped burn marks on the walls

   (d) All of the above

4. Lead azide and potassium perchlorate are examples of:

   (a) Detonator charges

   (b) Primary high explosives

   (c) Secondary high explosives

   (d) Both A and B

5. Which is an example of a medium petroleum distillate?

   (a) Butane

   (b) Gasoline

   (c) Charcoal lighter fluid

   (d) Diesel fuel

6. What are the four parts of the fire tetrahedron?

   (a) Heat, free radical reaction, fuel, carbon dioxide

   (b) Ignition, oxygen, fuel, primer

   (c) Chemical reaction, fuel, oxygen, heat

   (d) Heat, ignition, oxygen, retardant

7. Plastic explosive, dynamite, and PETN are examples of:

   (a) Primary high explosives

   (b) Secondary high explosives

   (c) Tertiary high explosives

   (d) Low explosives

8. Which of the following is not a high explosive?

    (a) Trinitrotoluene

    (b) RDX

    (c) C4

    (d) Firework propellant

9. An explosive is:

    (a) A material that undergoes a rapid exothermic reaction

    (b) A material that undergoes a rapid endothermic reaction

    (c) Both A and B above

    (d) None of the above

10. ANFO is an acronym for:

    (a) American North Forensic Organization

    (b) Aluminum nitrate ferris oxide

    (c) Ammonium nitrate fuel oil

    (d) Atomic Nuclear Fuel Organization

CHAPTER 11

Questioned Documents

Questioned document examination and analysis is an important and unique specialization in forensic science. The field of *questioned documents* covers a broad range of areas, central to which are the analysis of handwriting or printing and the examination of a physical document to prove or disprove authenticity. Questioned document examination is therefore a pure aspect of applied forensic science: the comparison of knowns to unknowns. Known elements may be handwriting or printing exemplars, or standards of paper, inks, dyes, and the like. The unknown elements, therefore, may be anything: ransom notes, forged checks, contested wills, threatening letters, counterfeit money, and of course the papers, inks, and dyes that are used to create the contested documents.

Sometimes, however, the end result of a questioned document examination is not measurable. Perhaps the inks from a letter can be run through a gas chromatograph-mass spectrometer (GC-MS) and matched to a known type of ink, but when it comes to the more detailed analysis of comparing and matching the strokes and flourishes of handwriting, there is no current way to measure the statistical accuracy of the results. Therefore, an expert in forensic document examination must be experienced and knowledgeable to make the proper interpretations of the results of the investigation. This level of reliance upon the expert's opinion is not unique to forensic document examiners: the same can be said for trace evidence examiners and hair examiners as well. The very definition of *expert witness* in all of the professions mentioned signifies that these individuals have more knowledge and specialization than the lay person. When an expert applies this specialized knowledge to the expert's particular field and makes a determination as to the results of a case, the result is not a simple guess. The result is built upon hours, if not days, of forensic examination, interpretation of the results of the observations, and patient analysis of all the details of the evidence. In the end, the educated elucidation, offered as the expert's opinion, is based not only on the science, but the experience and knowledge that the expert has attained as well.

Evidence that a forensic document examiner deals with may take the form of a handwritten note, a typed, printed, or copied form, or some other document. Questioned document analysis covers a broad range of document examination. An investigation can be done on the impressions, characteristics, and nuances of writing, printing, or copying, and in certain rare cases may also focus on the use of grammar, syntax, and punctuation. Analysis of the document may be as simple as a visual or microscopic examination of the makeup of the paper, or as in-depth as infrared (IR) spectroscopy analysis of the dyes in the ink, and may even include chemically treating the paper for the presence of other forms of impression evidence, such as fingerprints.

Questioned document examination is used in cases of forgery, fraud, and counterfeiting. It may also be important in kidnapping, threat, or suicide cases in which a note was found as part of the investigation. Civil cases also routinely rely on the expertise of a forensic document examiner, including for the authentication of wills and other legal documents and the verification of authenticity of historical documents.

Forensic document examiners are well trained in forensic science in order to understand and carry out the major task of identification and individualization. They should also understand pattern evidence interpretation and the role of comparison analysis in forensic science. The forensic background also comes in handy when dealing with the chemical and instrumental analysis of the inks, dyes, and paper, which is another major aspect of questioned document analysis.

Apart from forensics, anyone performing questioned document analysis must have very good eyesight, must not be colorblind, and must be comfortable looking through a microscope and magnifying glass for hours on end. Also, because one of the key elements of questioned document analysis is the analysis of the writing or phrasing of the document as a whole, a forensic document examiner should be well educated in grammar and linguistics (see the sidebar "Example of Native-Language Syntax").

Example of Native-Language Syntax

In questioned document analysis, there is not an exclusive focus on the English language. All languages are equally important, and knowing the nuances and phrasing of certain languages might come in handy in establishing a link between an individual and their writing style, whether it be in English, Farsi, Japanese, or French. For example, it is common in many European languages to put the subject at the beginning of a question (¿Marta, que vas a ir a la tienda? *Martha, are you going to go to the store?*), whereas in American English the subject in question usually comes later, if at all (Are you going to the store?). If a ransom note written in English is received, and questioned document analysis identifies notable differences from standard English in the syntax and diction, the native language of the author of the note might be discerned.

Certification

The courts determine if an examiner is qualified to give expert testimony in questioned document analysis cases. Some forensic document examiners choose to seek voluntary accreditation. One such accreditation is offered by the American Board of Forensic Document Examiners (ABFDE). To issue certification, the ABFDE reviews the education and training of the forensic document examiner, and verifies that the examiner has at least two years of full-time work experience in a questioned document examination laboratory under the supervision of a senior scientist. To be certified, the examiner also needs another two years of full-time independent, questioned document casework. The certification process for questioned document examination involves a written exam, a practical exam, and an oral exam. Only after successfully passing these exams can a forensic document examiner be certified as an expert. Certification is helpful for forensic document examiners because, when testifying in court, the determination of whether they are an expert is based on whether they have sufficient background and experience. Certification helps to establish the examiner's credentials to the judge and jury.

Tools of the Forensic Document Examiner

First and foremost, a good forensic document examiner must have excellent eyesight and the patience and endurance for careful and detailed analysis in the lab. While most people's vision problems can be easily corrected these days, the importance of the "eye for detail" in questioned document analysis is critical. The reason for this is that the evidence that is typically examined is not the handwriting itself, but the nuances, stroke direction, ink application, and even erasures that make up the minutiae of handwriting analysis. To get an idea of what is needed for questioned document analysis, one must realize that the typical minutiae being investigated are no bigger than the period at the end of this sentence. For other types of questioned documents, it is necessary to examine the fibers, inks, or chemical makeup of the paper or notes in question. Again, eyesight is an important factor. Apart from eyesight, though, the next most common tool for questioned document analysis is general magnification optics.

MAGNIFICATION OPTICS

The most common optical tool used in questioned document analysis is the magnifying glass, also known as a *lens* or *loupe*. This provides simple magnification of objects, not high magnification. Therefore, the magnifying lens is useful for a gross examination of the document or writing sample in question. This may also suffice for an experienced forensic document examiner to find an exclusion between a known and unknown sample, or to eliminate a possible suspect from having written a note, letter, or document.

To examine the fine details, a typical forensic document examiner then moves to a *stereomicroscope*. Typically, anywhere from 5x to 25x magnification is sufficient to examine handwriting, papers, and the like. To get an idea of 25x magnification, the period at the end of this sentence would be the size of a marble, and the letter *o* in "of" would be the size of a golf ball, and therefore the minutiae that make up these marks would be easily identifiable. Any higher magnification would be pointless for the purposes of examining the fine details in questioned document examination. The only possible exception to this is the examination of the fibers in paper to see the interaction of ink from writing or printing. For this examination, a magnification of 100x might be useful.

SPECTROSCOPIC ANALYSIS

Another useful tool in questioned document analysis is the ultraviolet-visible (UV-VIS) spectrophotometer for the process of UV analysis. Since inks and dyes have specific colors associated with them, these colors can be measured using UV-VIS spectrophotometry and then matched back to standards or compared against known ink samples to find a correlating color. While this is useful, many inks these days use roughly the same colors of dye to make up blue, black, and other regularly encountered colors in pens and markers. Moreover, these dyes are organic based, and therefore inexpensive, so they can be expected to be seen throughout many different types of evidence and across many different brands of inks. UV analysis is useful in identifying the "rare" colors that might be encountered in unique cases, which may provide a link back to a specific type or brand of ink if the color is unique enough.

A more common instrumental approach to questioned document analysis is still spectrophotometrically based, but uses infrared radiation instead. Although the colors of inks might be common, the various chemical makeup of one black or blue ink to another black or blue ink might vary widely. For this reason, IR analysis can be used to quickly differentiate between the two different inks. While this can be carried out on a Fourier transform IR (FTIR) instrument, it is more commonly done in questioned document analysis simply by using IR film and photography, or a special IR filter and a special digital camera.

Either way, IR imaging of a questioned document can quickly show differences in inks and ink types, as well as highlight erasures, chemical enhancements, washes, and other forms of manipulation to a document. For example, specific chemical solvents can be used to "erase" a person's signature from the back of a credit card. The signature area of a credit card these days is a unique area and is highly susceptible to smearing or blurring if rubbed or scraped, so the signature, once applied, cannot easily be erased with an eraser or scraped away with a knife. However, under the proper conditions, chemical solvents can be used to remove the ink from the original signature and leave the signature area intact. The forger can then apply his or her own style of signature to that line and use the credit card fraudulently. However, the chemical solvent interacts with the signature area and soaks into the background, leaving a telltale "watermark" that becomes visible when viewed using IR imaging. When this chemical mark is present, there is a good chance the card has been modified and the signature is a forgery. At the same time, IR illumination might also enable the examiner to "see" the remnants of the original signature, and help identify the true owner of the credit card.

Infrared imaging also comes in handy when analyzing checks and other financial instruments that might have been modified with two different inks—one to originally sign the document, and another to modify or forge it. The IR differences from one ink to the next, as mentioned before, might be enough to see the addition of a couple of zeros at the end of a dollar amount, the movement of a decimal point, or the modification of a payee's name.

CHEMICAL ANALYSIS

Another useful test in questioned document analysis is not instrument based at all, but rather comes from the science of fingerprinting. When paper is handled with bare hands, it acts like a sponge at a microscopic level, soaking up the oils, proteins, fatty acids, and cells that are present on your fingers. These oils and chemicals are what are typically left behind on other surfaces in the form of fingerprints, and on paper it is no different. However, since the oils tend to soak into the paper, simply dusting or fuming the paper will not readily produce any high-quality ridge detail on possible prints. Instead, a chemical called *ninhydrin* is used to treat the paper and develop any possible fingerprints that are present. Ninhydrin reacts with the amino acids in a fingerprint and turns from clear to a dark red/purple color in the presence of such amino acids. Ninhydrin is one of the few chemical-based developers that can be used successfully on porous surfaces. However, it can be damaging to the paper or document, so before ninhydrin is used, it is recommended to try less-harmful treatments like iodine fuming or the use of DFO. *DFO*, or *1,8-diazafluoren-9-one*, is another chemical developer that is meant to replace ninhydrin; it is less toxic and destructible than ninhydrin, but is also less sensitive.

INSTRUMENTAL ANALYSIS

For more in-depth analysis of the inks, dyes, paper, and chemical treatments of documents, a questioned documents examiner can turn to standard instrumental techniques for analysis and information. One of the common factors that is compared in questioned document analysis is ink and ink colors, which can be analyzed in various ways. As mentioned before, different inks are made up of several colors and components, and when these components are separated and examined, the ink may be identified and possibly even individualized. For example, a simple thin-layer chromatography (TLC) run can easily separate the components of inks and dyes, and allows the individual components to be isolated, extracted, and further analyzed. Using a GC-MS, it is possible to obtain a chromatogram and mass-to-charge analysis of the ink or dye, and compare this data to a library of known inks and dyes. If a manufacturer can be matched to the questioned ink or dye, the forensic analysis can then identify the type of pen or printer that was used to produce the document in question. That data point can aid investigators in further eliminating suspects and provide focus for the investigation.

Beyond the standard instrumental analysis, like GC-MS and TLC, specialized instruments are used by a forensic document examiner. One such instrument is the *electrostatic detection apparatus (ESDA)*. The ESDA is used to detect the indentations on a piece of paper that was below another sheet of paper that was written on. To do this, the sheet is subjected to an electrostatic field. Charged electrons from this static field are attracted to the damaged or impressed fibers in the paper where the indentations have been made and the underlying fibers have been stressed and mildly charged. Because these indentations are charged, a black toner or other type of charged particle can be deposited over the surface of the paper. The toner or charged particles are attracted to, and adhere to, those areas that have indentations. When the toner or charged particles are then fixed in place, the writing becomes pronounced and appears on the impressed page (see Figure 11-1).

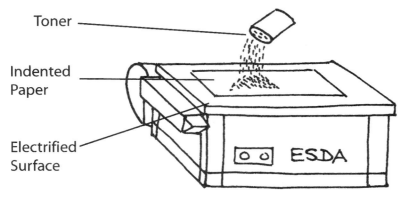

Toner

Indented
Paper

Electrified
Surface

Figure 11-1 An ESDA

This electrostatic discharge principle is the same that is used by a laser printer or photocopier to "print" on a page. An image of what is to appear on a page is electrostatically "burnt" into the surface of a metallic drum, which the paper is rolled over. At the same time, the opposite side of the paper is "washed" with toner. Toner, which is attracted to the charged areas on the drum but sticks to the paper instead, is then heated, which "melts" it into place on the page. When the paper exits the printer or copier, the image is fixed on the page.

The electrostatic component of the ESDA is also much like the static print lifter that is used in trace evidence examination for lifting dusty prints off of nonporous surfaces, such as shoeprints off of countertops or tire-tread marks off of a garage floor. The same principal is used, but instead of paper, a sheet of Mylar is placed over the dusty print and the metal sheet is electrostatically charged. The dust in the print is immediately attracted to the metallic surface, and, if protected or covered with a fixative, the print can then be further analyzed.

A more advanced instrument for questioned document analysis is the *video spectral comparator (VSC)*. The VSC is an advanced digital imaging device that uses a variety of oblique, direct, and transmitted illumination of various light sources, such as white light, UV light, and IR light. A digital camera and computer interface are capable of analyzing the spectral output of the document or inks in question, as well as providing a form of documentation about the analysis. Since such a broad range of light sources is used, a forensic document analyst can use a variety of physical and digital filters to isolate a single wavelength or a range of wavelengths to investigate a document. The VSC is also capable of digitizing some of the basic measurements of any writing or printing on a questioned document, and is capable of determining the angle between two lines, the circumference of a circle, or the area of a defined shape. This is important, because these factors are all typically measured by hand during the process of handwriting or printing analysis.

Types of Evidence in Questioned Document Analysis

It is important for a forensic document examiner to have a good understanding of the tools, techniques, and methods that are available for the purpose of questioned document analysis. All questioned document evidence can be broken down into two main categories, which are handwriting and printed documents. There are different reasons for performing the analysis on each of these evidence types, and aside from the common criminal element that is associated with forensic science, a forensic document examiner might also serve to authenticate a handwritten or printed document for legal reasons as well.

HANDWRITING

The practice of analyzing handwriting and printing is one of the most fundamental aspects of questioned document analysis. It is also one of the most subjective, and therefore one of the most difficult aspects of forensic document examination. Thus, a majority of the time spent in training and becoming qualified in questioned document examination revolves around this practice. The idea behind handwriting analysis is that the nuances and minutiae of a person's writing do not change even when the person is trying to "cover up" or forge a writing, signature, or document. Also, when a person is attempting to consciously change their writing, there are indicators in the writing that show the hesitation and "thought" that must go into the process. To uncover these factors, the actual microscopic aspects of the writing must be examined by the forensic document examiner.

In general, if a person is trying to copy another person's handwriting, the letters and lines have more stops and starts than there should be. For example, a person writing the capital letter *B* might use two total strokes, one to create the side of the *B*, and the other to create the two humps. However, if another person is trying to forge that *B*, they will probably make more strokes, possibly three or four, to complete the copy of the letter *B* (see Figure 11-2). Whereas the start and stop of each stroke might be indistinguishable to the naked eye, under a microscope, the strokes are easy to identify because they appear similar to overlapping strokes of paint on a canvas. In the preceding example, if the first person created the humps of the *B* with one stroke, then the ink line will change direction but not break between the two humps, whereas if the second person used two independent strokes to create each hump, the intersection of the humps will microscopically show overlapping, separate lines.

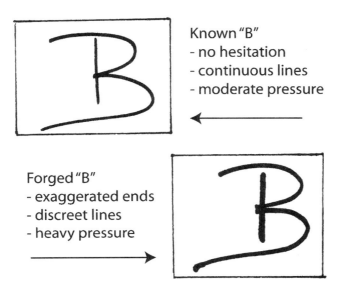

Known "B"
- no hesitation
- continuous lines
- moderate pressure

Forged "B"
- exaggerated ends
- discreet lines
- heavy pressure

Figure 11-2   Examples of a known *B* and a forged *B*

Another key factor in handwriting is that the flourishes and slant of the writing are difficult to modify. Most right-handed people tend to write with a slant to the letters, such that the top of the letters point to the upper right, and the bottom of the letters point to the lower left. The opposite is true for left-handed people; their writing tends to slant in the upper-left to lower-right direction. Therefore, when a right-handed person is attempting to copy left-handed writing, the actual slant of the letters must be reversed as well. Unfortunately for forgers, writing is a deeply engrained process. We have been taught to write at a young age, and as we grow our hands and fingers develop in specific, unique ways. As a result, our writing style becomes "fixed" in our physical structure, and it is always more comfortable to revert to that style, or aspects of that style, when writing. To force an opposite slant when forging, the writing might appear slow and show signs of hesitation, because the brain is actively working against instinct. These *hesitation* marks may show up as pronounced wiggles in the body of the letters or as unnecessary flourishes in the actual writing.

While overly used or grand flourishes may be a sign of forgery, they are not a sign of poor writing. Everybody who writes also develops specific flourishes in his or her own writing style. A flourish may be as simple as the way one dots a lower-case *i*, to something as complex as the underline on an official signature. Whatever the flourish, it is again one of the normal aspects of handwriting and it is also engrained in our standard writing. Even when trying to forge a document, then, a forger may be paying attention to the gross details of the handwriting, such as the spacing, size, and slant of the letters, but still dotting *i*'s and crossing *t*'s using his or her standard method, and not the normal flourishes of the person whose handwriting is being copied.

One of the final aspects of handwriting analysis is the comparison of the overall size, shape, and area of certain letters that a person makes. Not all of us write in a specific font size, nor do we all close our *O*'s or create the circumference of our *C*'s the same way. Careful measurement of these factors can aid a forensic document examiner in determining a forgery from an authentic document. But, as in all types of forensic science, this analysis is not done blindly. For any final determination of forgery, a forensic document examiner must obtain *exemplar writing samples* from the suspect and the victim.

Obtaining these writing samples from the victim is usually fairly straightforward. Having nothing to hide and everything to gain from the investigation, a victim of forgery usually provides a straightforward writing sample. The suspect in a forgery case, on the other hand, must also provide an exemplar writing sample. Because such a person might face arrest and jail time if proven to be a forger, they are under pressure, if guilty, to again forge their writing. Therefore, during the collection of the writing sample, the same strategy is employed for both the victim and the suspect in the case. Each person is typically asked to submit a writing sample that includes various forms of printing, script writing, and block writing, which also should include every letter and number in the alphabet. A common phrase like "A quick brown fox jumped over the lazy dog" uses every letter in the English alphabet, and when a guilty person is forced to write this in different manners over and over, they will begin to lose focus on any attempt to forge their writing. In doing so, they will revert to their standard style if they are attempting to cover up their writing.

When this exemplar handwriting sample is then examined by the forensic document examiner, a trained eye can spot areas in the writing sample where alterations in the writing style changed and to what degree it changed as the person continued to write. If comparison of the flourishes, stroke and stroke pattern, and similar letter formation reveals distinct similarities between the suspect's writing and the forgery, the forensic document examiner may rule it a match, or that the person could not be excluded as a contributor to the forgery.

PRINTED DOCUMENTS

This same type of comparison can also be done on mechanically printed forms, and even though the unique element of the human hand is removed from the creation of these forms, the analysis of a known and unknown printing can be even more difficult than handwriting comparison. To begin with, the same comparative principles still apply; the questioned unknown document that contains printing also contains the flourishes and uniqueness of the printer. Such imperfections used to be much clearer in the past, when typewriters were used to create printed forms. As the typeface pressed the inked ribbon into the paper surface, the minor imperfections in the typeface surface were transmitted to the inked print on the paper. Such pits, mars, and scratches on the typeface surface could therefore be identified on the typed form, and with careful analysis, the original machine could be identified. However, typewriters are used infrequently these days, having been replaced with inkjet and laser printers, photocopiers, and thermal imaging machines. Such printing on these machines does not utilize a regular type of impression interface as in typing, but instead relies on ionic forces (see the earlier discussion of ESDA), microjets, or piezoelectric heating.

Despite the lack of impression evidence on modern printouts, there is still a large amount of individual data that is microscopically present on most printed forms. In the cases of printers that use a metal drum for imaging, any dirt, debris, nick, or gouge in the metal can result in a unique and reproducible artifact on each and every printout. This is commonly seen in laser printers, photocopiers, and thermal imagers as black smudges or blots on the printed surface. Because these are inherent defects in the printer system, they will be seen over and over again until the machine is repaired or the imaging drum is replaced.

For inkjet printers, defects in the print head can result in smears, lines, or voids in the printed image. If the print head is not cleaned or replaced, these defects will be continuously produced as more and more pages are printed. Also, specific defects in a batch of ink will affect all of the pages that are printed with that unique problematic batch. It is important to note that all of these identifiable defects can diminish or even disappear over time. Therefore, it is useful to obtain exemplar copies of printouts from machines as soon as possible. If one were to go back and look at the printouts from a laser printer from seven years ago, when it was brand new, there probably would not be a single defect on the printed page. Today, there are probably many smudges, spots, and artifacts. If these two pages are compared side by side, it might be extremely difficult, if not impossible, to prove that they came from the same printer. (See the sidebar "The Inkjet Printer Code.")

The Inkjet Printer Code

Interestingly, in 2005, it was revealed that some printers did more than just print images and text. A private group investigating civil rights violations discovered that certain inkjet and color laser printers applied a small series of yellow dots to all pages during the printing process. These yellow dots, when microscopically viewed and decoded, indicated the serial number of the originating printer as well as the date and time of the printout. Since serial numbers can be quickly traced to retail sales and therefore to individuals, this "hidden" system has the ability to uncover the origins of illegal printing, most notably counterfeiting. When this news was released, the printer company confirmed that the system of dots was in fact known as the *Counterfeit Deterrence System (CDS)* and was instituted by the Federal Reserve Board. This system also has a myriad of uses for forensic analysis, such as tracing ransom printouts back to a source and providing a timestamp on the printout without having to "date" the paper or the ink.

AUTHENTICATION

One of the final types of evidence that can be encountered in a questioned documents lab are legal or official documents, for the purposes of authentication. While not truly a forensic investigation, the authentication of a document uses the same principals of comparative handwriting analysis, chemical composition analysis of the paper and inks, and review of the grammar and syntax in the writing to prove or disprove the official origin of such documents. This type of work is typically done in cases of contested wills, deeds, and other questioned legal forms, as well as to authenticate famous signatures on memorabilia and other mementos. If a document, signature, or work is found to be authentic, the forensic document examiner may issue a certificate of authenticity for the item that summarizes the analysis and the final results.

Methods of Questioned Document Investigation

One of the most useful methods of establishing whether a document is authentic or forged is by *dating* the materials. This can be done by instrumental and microscopic analysis of the inks, dyes, and paper used to create a document. If, for example, a "lost" testament of the Bible were uncovered in the hills outside Jerusalem, analysis might reveal that the parchment contains bleached, recycled plastic fibers, and that the writing itself was made using a black crayon. Since neither of these components actually existed in biblical times, it would be fairly easy to identify this document as a fake. Obviously, then, this type of analysis requires a specialized knowledge of inks, paper, and dyes of certain time periods. For this type of information, forensic document examiners turn to archives, production logs, databases, and indices that detail the broad history of such items. Based on this data, a forensic document examiner can readily determine, for example, when a company incorporated a specific dye set into an ink, or stopped using certain fibers in paper production.

PAPER

When it comes to the analysis of paper, one should consider the numerous types of paper that are commercially available today. What tends to be a common factor about most paper, especially the typical 8½×11-inch paper that is fed into most printers and photocopiers, is that it is universally produced, usually from recycled content. The same goes for newspaper. Paper is typically made from wood pulp that has been processed, bleached, dried, pressed, and cut to form pages. If cotton or linen fibers are added to this pulp, the resulting paper is known as *bonded* or *rag* paper. These bonded or rag papers usually contain *watermarks*, which are distinctive marks left in the paper by the manufacturer that are visible when the paper is held up to light. These bonded or rag papers are typically used for special or important documents, such as résumés and wedding invitations, and are therefore not as common. However, ironically, it is the addition of the cotton, linen, or other special fibers and the presence of a watermark that make it easier to identify and individualize a particular paper when examined as evidence. It is next to impossible to even identify the manufacturer of the common recycled printer paper, due to the lack of such specialized components.

INKS AND DYES

Ink is another commonly analyzed component in questioned document analysis. By extracting and analyzing ink from a document, the individual components, dyes, and additives can be identified and compared back to manufacturer databases and indices. Traditional inks used in high-end pens might be similar to India ink, which is a mixture of black carbon in a binder solution. Certain fountain pens used to use another type of ink known as iron gallotannic (or iron gallnut) ink, and while some still use it to this day, most have modernized. Today, most pens and markers use inks that are based on organic chemicals, called *dye inks*. Dye inks are inexpensive and much easier to produce than the older-style inks, and thanks to modern chemical techniques, dye inks can be made in any color imaginable. Almost every inkjet printer and cheap ballpoint pen uses some form of organic dye ink.

When ink is deposited onto paper, it is applied in a wet form and allowed to dry. The drying process is actually the evaporation of the binder or liquid medium in which the actual dye was suspended. In inkjet printer inks, the ink is designed to be attracted to the cellulose structure of the paper, allowing the dye to penetrate the paper and the binder to eventually evaporate. This is usually a rapid process, since most inkjets deposit only micro-droplets of ink onto a printed page. For pen ink, the volume of ink dispensed is much larger, and it takes a longer amount of time for pen ink to dry compared to inkjet ink.

Adulteration, Erasures, and Additions

When examining the aspects of writing on a document, a forensic document examiner is looking not only for the unique features of the writing, but also at what is missing or out of place. Any modification to writing, especially in terms of forgery, is known as *adulteration*. Adulteration is the process in which the writing is erased, eradicated, or modified to achieve the expected outcome. Adulteration can be as simple as erasing one signature to apply another or as complex as manipulating the ink on the page on a microscopic scale to change the writing on the page.

Unfortunately for the forger, the basics of forensic science (such as Locard's Exchange Principle; see Chapter 1) still apply, and every contact leaves a trace. In questioned document analysis, this contact is twofold. First, the original writing leaves trace impressions in the paper's surface, and as ink or some other writing medium is deposited, the underlying fibers on the page are impressed and "damaged." Second, the adulteration of the first writing leaves behind telltale marks as well. The following is a short list of possible adulterations that are commonly seen in questioned document analysis:

- **Manual erasure**   The fibers on the page show the back-and-forth stroke of the eraser surface across the paper fibers, and usually microscopic amounts of the eraser and the original ink are left behind.
- **Scraping**   The forger uses a razor to scrape away the ink, causing the underlying fibers to show damage from the blade at the microscopic level.
- **Chemical erasure**   Solvents are used to remove the ink, and these chemicals leave behind telltale watermarks and trace amounts of the chemical that can be detected with basic instrumental analysis.

In any of the preceding methods, the new writing also leaves new impressions in the paper's surface and redeposits of ink, which can be used to identify not only the forgery, but the tools used to execute the fraudulent activity as well.

When adulterations are made using additions to the existing text, the forensic analysis of the evidence is performed in a different way. Instead of looking for what was removed, the forensic document examiner looks instead to find what was added. Sometimes, the addition can be simple. For example, if the decimal point on a check for $100.00 is changed to a comma and an extra zero is added, it becomes $100,000. The writing on the check would also need to be modified, but the simple addition of a small line to a period and an extra zero changes the amount of a check dramatically.

Additions are detected microscopically, such as looking for the beginning of the new stroke lines in the added text, and the differences in the two types of ink used. Because additions are usually done with a pen, marker, or pencil of the same color, on gross examination the inks might be the same color. Subject the adulterated document to UV or IR light, however, and if the two inks have different chemical properties, the differences will readily be apparent under the alternate light source.

Have you ever wondered why, when paying certain bills, certain companies specify the long form of their name to be used on checks? For example, when paying income taxes to the federal government, the IRS specifically says "Please make out checks or money orders to the Internal Revenue Service." The reason for this is the long history of additions that have been made to checks when abbreviations, such as IRS, were used in the original. Imagine what could happen if you only wrote "IRS" on the payee line of a check and mailed it off. Anybody who illicitly recovered that check could easily add a couple of lines, a period, and a name to that title, creating a check to "MRS. SMITH" (see Figure 11-3). They could change the *I* in IRS to an *M*, add a period and a name, and cash your check—not only do your taxes not get paid, but somebody posing as Mrs. Smith now has your money!

```
┌──────────────────────────────────────────────────────────────┐
│  Jane Doe                                     5301             │
│  110 Any Ave                                                  │
│  My City, NY 00235          June 3, 2005                      │
│                                       DATE                    │
│    PAY TO THE                                                 │
│    ORDER OF    IRS                        $ │1232.00│         │
│     One thousand two hundred thirty-two ——— DOLLARS           │
│                                                               │
│    MEMO                        Jane Doe                       │
│                                SIGNATURE                      │
│   986543513   541698350  5301                                 │
└──────────────────────────────────────────────────────────────┘
```

```
┌──────────────────────────────────────────────────────────────┐
│  Jane Doe                                     5301             │
│  110 Any Ave                                                  │
│  My City, NY 00235          June 3, 2005                      │
│                                       DATE                    │
│    PAY TO THE                                                 │
│    ORDER OF    MRS. SMITH                 $ │1232.00│         │
│     One thousand two hundred thirty-two ——— DOLLARS           │
│                                                               │
│    MEMO                        Jane Doe                       │
│                                SIGNATURE                      │
│   986543513   541698350  5301                                 │
└──────────────────────────────────────────────────────────────┘
```

Figure 11-3 An example of converting IRS to MRS. SMITH

Counterfeiting and Countermeasures

One of the most commonly forged documents is paper money. This process is known as *counterfeiting*. With the recent advancements in digital imaging technology, such as high-quality color scanners and inexpensive color printers, it is getting easier and easier to counterfeit money. That is, it *was* getting easier. Recently, the U.S. Bureau of Engraving and Printing started producing new versions of common currency. While these new changes are designed to thwart new forms of counterfeiting, they only make sense if you understand the old process in which money was produced.

Currency in the United States is printed on special paper, produced exclusively by Crane & Company, a small firm that also produces high-end stationery and other forms of bonded paper. The paper used in currency is produced under protected conditions and is only available to the U.S. Government Printing Office. This paper also has special properties that are used to thwart counterfeiting. One of the main aspects of the paper itself is that it is infused with incredibly small colored threads, usually red and blue. If you hold up any denomination and look at it closely, these threads are small, about 1 to 2mm long, and appear to be debris on the background. These threads are added to the paper pulp when it is mixed, and they occur in a specific concentration and ratio of colors on different denominations.

The color of money is also an important anti-counterfeiting measure. In the past, a special secret blend of dyes was used to create the various green hues on the printed money. Now, as described shortly, multicolored ink is being used. The final traditional anti-counterfeiting measure is known as a security thread. This thin strip, about 3 to 4mm wide, is embedded in one random location running from top to bottom in every bill larger than $1. This strip itself has printing on it, visible through the paper when held up to light, that contains the denomination of the bill; so the strip in a $5 bill has the word "five" printed over and over, and a $20 security strip has the word "twenty" printed over and over.

The newest forms of currency still use the paper with the colored fibers and the security thread, but the traditional green ink is replaced with colored ink that is applied to give a layered, textured look to the money. The printing of this color is done in various steps and in various degrees of density across the currency, so it is very hard to copy. Also, the way in which the printing is done produces very fine lines in complex and shifting patterns, which are easily recognized by a human eye but are difficult to resolve using modern scanners or digital imaging equipment. Likewise, microprinting across the bill's surface with UV ink has been added. This is done so that, when held under a UV light, tiny 20s, 50s, or 10s, depending on the denomination of the bill, are visible across the background of the money.

Another special ink, known as *color-shifting ink*, is also used on the newest currency, usually on the numerical denomination in the lower-right corner of each bill. This color-shifting ink "changes" colors as the bill is tilted back and forth, so that the 20 in the lower-right corner on a new bill shifts from bright green to dark black.

All of these countermeasures were designed with specific goals in mind, not only the effectiveness of the anti-counterfeiting measures. The new currency is rugged and durable and capable of being wet or washed and still viable for use. Also, even though colors and different, larger images are being used, the money still has an "American" look and feel. Also, as a result of the larger images and bolder print on the bills, the new currency is easier to distinguish by the elderly or those with limited eyesight.

While all of these features are important and designed to thwart counterfeiting, there might come a day when digital technology has the ability to copy and print each and every one of these surface features. However, one of the most important security features on currency cannot be copied, because it is truly *not there*. The watermark in the newest currency is one of the most important counterfeiting countermeasures. The watermark for each denomination is different, and is a replica of the portrait that appears on the bill. The watermark is imprinted onto the raw paper during the process in which the pulp and rag content is still wet and pressed flat. This feature is only then visible after the paper dries by holding up the bill to the light. This feature cannot be duplicated by a copier or a printer, because it is impossible to copy or print what is invisible to the naked eye!

Quiz

Refer to the text in this chapter if necessary. Answers are located in the back of the book.

1. In handwriting analysis, common characteristics that might mistakenly appear to be forged are:

    (a) The gross features of the handwriting

    (b) The flourishes of the handwriting

    (c) Punctuation

    (d) Letter forms

2. The most common ink found in inkjet printers is:

   (a) India ink

   (b) Nigrosine ink

   (c) Organic dye ink

   (d) None of the above

3. Which of the following is the least destructive method in document examination?

   (a) GC-MS

   (b) Infrared spectroscopy

   (c) TLC

   (d) Infrared photography

4. Which of the following is an individualizing feature of typewriters?

   (a) Ribbon tension

   (b) Cavities on the typeface

   (c) Font

   (d) Spacing of letters

5. Which of the following is not an individual characteristic of handwriting?

   (a) Spacing

   (b) Phraseology

   (c) Margins

   (d) None of the above

6. Individualizing marks seen on photocopies originate from:

   (a) The glass

   (b) The original copy

   (c) The drum

   (d) All of the above

7. India ink is commonly made of what?

   (a) Tar and solvent

   (b) Carbon and binder solution

   (c) Organo-phosphate dye in water

   (d) Hematoxylin and potassium chromate

8. Which of the following is not a sign of possible deception in questioned document analysis?

   (a) Retouching of letters

   (b) Overemphasis of initial strokes

   (c) Wavering lines

   (d) Legibility

9. What is not a characteristic of rag paper?

   (a) Watermark

   (b) Damaged fibers of a distinctive length

   (c) Made entirely of wood pulp

   (d) Both B and C

10. What anti-counterfeiting measure in U.S. currency cannot be duplicated in any way with current digital technology?

   (a) The watermark

   (b) Security threads

   (c) Microprinting

   (d) The special paper

CHAPTER 12

Forensic Psychiatry

By David Taylor, M.D.

As explained in Chapter 1, the word *forensic* means "matters pertaining to law." A *forensic scientist* is someone who applies his or her scientific knowledge to legal questions. Most professional disciplines have both forensic and nonforensic sub-specialties. For example, a general dentist cleans your teeth and fills your cavities, while a *forensic odontologist* might be asked by the court to identify a cadaver based on dental records. In both cases, the dentists make use of their knowledge of dentistry, but only the forensic dentist applies that knowledge to legal matters.

Psychiatry is the branch of medicine that deals with the diagnosis and treatment of mental illness. Like many other scientific fields, it too has a forensic subspecialty. This chapter focuses on forensic psychiatry. It begins with some general information, addressing such questions as: What is mental illness? How are psychiatric diagnoses made? What does a forensic psychiatrist do? Then the chapter introduces the legal system and the history of psychiatry to put things in perspective.

Finally, a number of common legal issues that involve forensic psychiatry are considered: What does it mean to be not guilty by reason of insanity? Does a defendant have to be mentally competent to stand trial? Do all sex offenders have mental illness? Can people with dementia write a will? If an employee is injured at work, is the company responsible? Can judges terminate parents' right to raise their children? How do you negotiate with a terrorist? Along the way, several case examples will also be presented.

A Brief History of Psychiatry

The treatment of the mentally ill is at least as old as the Bible. The Old Testament describes a depressed King Saul tormented by an "evil spirit" and soothed by David's harp (I Samuel, Chapter 16). Physicians, however, did not specialize in mental disorders until the 19th century, which saw the rise of exclusive treatment facilities called *asylums*, the precursors to modern psychiatric hospitals.

Little was known about the causes of mental illness at the time, and many people believed that these diseases were the result of divine punishment, ethical shortcomings (called *moral insanity*), or imbalance of body fluids (sometimes treated by *bleeding*). Gradually, physicians began to suspect the brain as a cause of impairment. Dr. Franz Gall (1758–1828) pioneered the field of *phrenology* in which the contours of patients' scalps were studied to diagnose problems of underlying "brain organs" (see Figure 12-1.) Although Dr. Gall correctly identified the brain as the source of mental disorders, the field of phrenology has since been discredited and shown to have no scientific merit.

Without a clear understanding of the physical causes of disease, physicians began to explore new ideas about mental illness. Some doctors suspected that mental disorders resulted from psychological causes. The most famous of these physicians, Sigmund Freud (1856–1939), developed a theory of personality based on unconscious sexual and aggressive drives. Freud also invented a treatment for these problems that he named *psychoanalysis* or "the talking cure."

Over the following decades, much has been learned about diseases of the mind, both their causes and their treatments. Rotating chairs, cold baths, and insulin comas have been replaced with psychopharmacology, psychotherapy, and somatic treatments. Let's turn now to how mental illness is understood and diagnosed today.

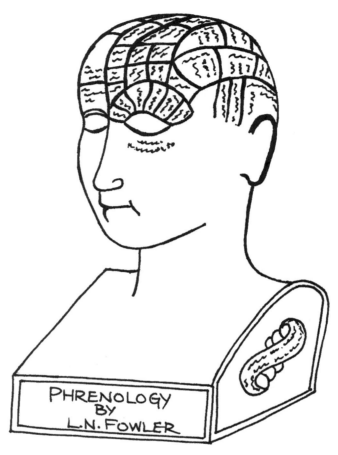

Figure 12-1   Phrenology bust

Mental Illness Today

Mental illness is exceedingly common. The National Institute of Mental Health (NIMH) estimates that more than one in four Americans suffer from a mental illness during any given year. Not only is mental illness widespread, it also causes considerable impairment. One way that researchers compare the harm caused by different diseases is to analyze their relative impacts on the economy. Using this method, the World Health Organization's *Global Burden of Disease Project* found that mental illness contributes to more of the world's disease burden than all cancer cases combined.

When diagnosing mental illness, psychiatrists in the United States rely on a book called the *Diagnostic and Statistical Manual of Mental Disorders*, known as the

DSM for short. The first edition of the DSM was published in 1952 to provide a coherent classification scheme for mental illness. It organizes mental disorders into several general categories, such as mood disorders and eating disorders. Within each category, individual illness are described and compared—anorexia nervosa and bulimia nervosa, for example—and the defining signs and symptoms of each are listed. Since people express the same disorder in different ways, only some of the symptoms are needed in order for a psychiatrist to make the diagnosis. For example, the DSM lists nine possible symptoms of a major depressive episode but only requires five to make the diagnosis. This formal diagnostic procedure is essential because medical researchers have not yet been able to determine the precise cause of most mental disorders or to create laboratory tests that reliably diagnose them.

Before we explore some of the common psychiatric diagnoses, let's first take a step back and consider a fundamental question: what is mental illness? The DSM defines a mental disorder as a clinically significant syndrome associated with distress or impairment that is caused by behavioral, psychological or biological dysfunction. To be defined as a disorder, a syndrome must also fall outside the range of an individual's cultural expectations. For example, some religions consider seeing and hearing the dead a natural part of mourning, while others would consider such behavior abnormal. Although all of these conditions are called *mental disorders*, the name is somewhat misleading, as it seems to imply an artificial distinction between "mental" disease and "physical" disease. Today, mental illnesses are understood in a comprehensive biological, psychological, and social framework, a perspective known as the *biopsychosocial model*.

Some Common Mental Disorders

The DSM identifies hundreds of distinct mental disorders, each of which can have numerous descriptive qualifiers. In this section, we focus on a few of the common symptoms and disorders, including those that frequently occur in forensic contexts.

One of the most dramatic psychiatric symptoms is *psychosis*, or severely impaired reality testing. A common psychotic symptom is *auditory hallucinations*, which is the experience of hearing voices when alone. Another example of a psychotic symptom is having a *delusion*. A delusion is a persistent, false belief.

Often a psychotic individual has multiple symptoms at the same time. For instance, a person might hear voices whispering to him (auditory hallucinations) and believe those voices belong to FBI agents who are secretly spying on him (a delusion). *Schizophrenia* is a severe psychiatric disorder in which a person experiences several psychotic symptoms over many months. Despite popular notions, schizophrenia is unrelated to having "split personalities" (a condition known as *dissociative identity disorder*).

In severe psychosis, the individual's thoughts, speech, and behavior become completely incoherent. The person may act in an unexpected and bizarre manner. This level of gross disorganization often impairs the person's ability to cooperate with others. If such an individual were to commit a crime, a forensic psychiatrist might be asked by the court to determine if the defendant is mentally competent to proceed with the trial. (You'll read more about competency to stand trial later in this chapter.)

Case Example 1

A homeless woman with severe psychotic symptoms hears a voice telling her that she owns Bloomingdale's. She enters the department store, takes some merchandise, and leaves. When arrested for shoplifting, she insists that she did not steal because everything in the store belongs to her.

Questions for the forensic psychiatrist:

- Which psychotic symptoms does the woman demonstrate? What is her diagnosis?
- Does she appreciate that she took something from the store?
- Is she guilty of shoplifting? Should she be held responsible?

Another common mental disorder that psychiatrists frequently diagnose is *major depressive disorder*. It is normal to experience some variations in mood. Major depression, however, is different from ordinary blues. In this condition, a person experiences an incapacitating sadness every day for at least two weeks that prevents normal activities, such as eating, sleeping, and concentrating. Many depressed people have thoughts of suicide, and up to 15 percent of people with major depression kill themselves. Severe episodes of depression can even involve psychotic symptoms in which an individual might have delusions of being cursed or of deserving punishment.

Forensic psychiatrists are often asked to evaluate individuals for PTSD, or *posttraumatic stress disorder*. PTSD can be thought of as an understandable reaction to a traumatic event. The disorder was first recognized in Vietnam soldiers who, returning from combat, showed symptoms of nightmares, flashbacks, and increased vigilance. Lawsuits involving personal injury often claim psychological damages—including symptoms of PTSD—which must be assessed by a forensic psychiatrist.

Interestingly, one condition that forensic psychiatrists are frequently asked to assess is not a mental disorder at all, although it is included in the DSM. *Malingering* is defined as the intentional production of symptoms, in other words, faking mental illness. Most malingering occurs in order to achieve a specific purpose such as avoiding military duty or criminal prosecution. The assessment of malingering can be one of the most challenging aspects of forensic psychiatry because the diagnosis of mental disorders mainly relies upon the self-report of subjective symptoms. In general, forensic psychiatrists use collateral information and various psychological tests to confirm their opinions.

How Do I Become a Forensic Psychiatrist? Where Does a Forensic Psychiatrist Work?

Standing at the intersection of the medical and legal systems, forensic psychiatry is an exciting and fascinating field. It offers a unique opportunity to apply medical knowledge to legal questions and, in doing so, to address many complex social and ethical issues. As legal consultants, forensic psychiatrists help educate the public about mental illness and reduce the stigma associated with these serious disorders. As clinicians, forensic psychiatrists serve some of society's most vulnerable and marginalized individuals and improve the mental health of those who may need it most.

A forensic psychiatrist is a psychiatrist with specialty training in forensics. To become a general psychiatrist, one must complete many years of education: four years of college, four years of medical school, and four years of residency. To specialize in forensic psychiatry, one must then take an additional year of training called a *forensic psychiatry fellowship*. After completing all of this training, graduates may take a standardized exam given by the American Board of Psychiatry and Neurology to become *board certified*.

A forensic psychiatrist can work in a variety of settings. He or she may treat the mental disorders of prison inmates, a career also known as *correctional psychiatry*. A forensic psychiatrist might also assist law enforcement by doing such things as serial killer profiling or terrorist negotiations. Additionally, a forensic psychiatrist can perform evaluations for the legal system. When asked by a judge or lawyer to do an assessment, a psychiatrist spends many hours reviewing records, completing an interview, and writing a detailed report. Afterwards a psychiatrist might then be called into court as an *expert witness* to provide testimony. This testimony merely reflects the physician's professional opinion. It is up to the judge or jury to decide the case and issue a verdict. (See the upcoming section "An Introduction to the Law" for more information on expert witnesses.)

A Brief History of Forensic Psychiatry

It has long been accepted in English law that the mentally ill should not be punished for their crimes. Courts assumed that the determination of insanity was a social, not medical, issue, and doctors were rarely called upon to offer an opinion. The first forensic psychiatrists to give courtroom testimony did so in the late 1700s and were known as "mad-doctors." By the middle of the 19th century, forensic psychiatrists were regularly called upon to assess defendants. The increased role of forensic psychiatrists coincided with the growth of asylums in which physicians began to have increasing experience observing and treating the mentally ill.

The first well-known American forensic psychiatrist was Dr. Isaac Ray, who, in 1838, published *A Treatise on the Medical Jurisprudence of Insanity*. This textbook of forensic psychiatry was used around the world, and judges into the middle of the 20th century cited it in their decisions. Ray later became president of the first American medical organization, a group that eventually developed into the *American Psychiatric Association (APA)*. In its early years, APA membership consisted of asylum directors, many of whom gave courtroom testimony and considered forensic psychiatry to be one of the main functions of this new specialty.

In recent years, the role of psychiatrists in the legal system has come under increasing scrutiny. Public outrage over the violent and heinous crimes of a few has led the demand for stricter punishments. The verdict "not guilty by reason of insanity" has sometimes given the false impression that individuals are excused for their actions. Forensic psychiatrists themselves are at times uneasy with the scientific and ethical bases of their decisions. In the remainder of this chapter, we look at some of the many ways in which forensic psychiatrists today are asked to provide professional opinions on legal matters.

An Introduction to the Law

As mentioned in Chapter 2, with the exception of a few types of cases, most American courts are structured in what is known as an *adversarial system*. In this model, each opposing party retains a lawyer who presents to the judge or jury as favorable a version of the truth for their client as possible. The philosophy behind this approach is that the truth will be made apparent to the court by each party's best efforts to portray the most persuasive, honest position.

Case Example 2

A man with a long history of mental illness attempts to kill himself and his two young children after his wife divorces him. The man survives but the two children do not. The defense attorney hires a forensic psychiatrist to interview his client. The defendant tells the psychiatrist that, for two weeks before the crime, he heard the voice of God commanding him to kill himself and his children. The doctor testifies that the man was suffering from major depression with psychotic features and should be found not guilty by reason of insanity. The district attorney hires another forensic psychiatrist, who also interviews the defendant. This psychiatrist determines that the man did not have any psychotic symptoms at the time and murdered his children to punish his wife for the divorce. He testifies that the defendant is *malingering* his psychotic symptoms.

Questions for the jury:

- Which version do you believe?
- How do you evaluate the opposing opinions of the two expert witnesses?

There are two broad types of legislation: criminal and civil. In *criminal law*, the state prosecutes a defendant for criminal behavior. The state's attorney must convince a jury *beyond a reasonable doubt* of the defendant's guilt. The goal of criminal law is to punish or reform offenders and deter future crime. The other general type of legislation is *civil law*, in which one citizen sues another to resolve a dispute. The goal of civil law is to provide monetary compensation or injunctive relief to the damaged party. Unlike criminal law, blameworthiness is not necessary, and a jury must only be convinced by *a preponderance of the evidence*—that is, with more certainty than doubt—that the harm occurred.

There are several types of cases that can involve forensic psychiatrists. In the remainder of this chapter, we explore in greater detail some of the most common examples. The types of criminal cases that we will cover involve the insanity defense, competency to stand trial, and sex offenders. For civil law, we will consider the topics of testamentary capacity, workers' compensation, and termination of parental rights.

THE INSANITY DEFENSE

Forensic psychiatrists are frequently asked to perform a *sanity evaluation* by defense attorneys. In this evaluation, the physician determines if the defendant was legally insane at the time of the crime. Each state defines "legally insane" differently, but most require that the person was unable to appreciate the wrongfulness of his or her actions because of a mental disorder. Most courts permit severe mental disorders—

such as schizophrenia, major depressive disorder, and mental retardation—to qualify for this definition. Other conditions, such as alcohol intoxication, often do not meet the legal criteria.

If the defendant is found not guilty by reason of insanity, he or she is acquitted of the crime. As the verdict indicates, the defendant is "not guilty" and therefore cannot be punished for the offense. However, very few of the acquitted are immediately released. Because of their mental illness and recent dangerous behavior, many of these individuals are instead committed to forensic psychiatric hospitals where their disorders are treated and their dangerousness is monitored. Patients are released from the hospital when they are no longer considered mentally ill or dangerous to themselves or society.

Contrary to popular belief, the insanity defense is rarely raised in court. In fact, only about 1 percent of defendants charged with a felony plead insanity. Of these people, less than a quarter of them are eventually determined to be legally insane. Individuals may be reluctant to use the insanity defense because of the social stigma associated with mental illness or because of concern that, if found insane, they could potentially spend more time confined in a hospital than they would spend in prison if sentenced.

Case Example 3

A mentally retarded man robs a convenience store, kills the store clerk, and flees the scene. Upon arrest, a blood alcohol level shows the man was intoxicated at the time of the offense. Psychological tests confirm that the defendant's IQ is in the mild mental retardation range. A forensic psychiatrist is consulted to perform a sanity evaluation.

Questions for the forensic psychiatrist:

- Does the fact that the defendant left the crime scene mean he knew his actions were wrong? If so, was he therefore legally sane at the time of the crime?
- Can his impaired intelligence or alcohol intoxication be considered a mental disorder according to the legal definition of insanity?

COMPETENCY TO STAND TRIAL

Another criminal issue that forensic psychiatrists are frequently asked to assess is whether a defendant is competent to stand trial. Any party involved in a case—the defense, prosecution, or judge—can request a competency evaluation. These protections ensure the fundamental fairness and dignity of the trial.

A sanity evaluation investigates a defendant's previous mental state as it existed during the alleged crime. In contrast, an evaluation for competency is based on the individual's functioning at the present time. To be competent for trial, the defendant must understand the court proceedings and be able to assist in the defense. For example, he or she must understand the role of the jury and the nature of the charges made. If a mental disorder impairs these abilities, the defendant is determined to be incompetent to stand trial. The individual can then be transferred to a forensic psychiatric hospital in order to restore competency. Competency restoration is achieved by treating the underlying mental illness or by educating the defendant about court proceedings. Certain conditions, such as severe mental retardation, may not be amenable to treatment. The U.S. Supreme Court has ruled that individuals may not be held in a hospital longer than a "reasonable period of time" to determine if competency is likely to be restored.

Case Example 4

A man has a well-documented history of bipolar disorder, a mental illness in which symptoms typically relapse and remit. During a symptom-free period, the man robs a bank and escapes. Three months later he is apprehended after a routine traffic violation. The arresting officer reports that the man was driving erratically and ran a red light. When asked for his license and registration, the man replied, "I am the light and the lord! Heaven waits for no man."

Questions for the forensic psychiatrist:

- Is he competent to stand trial?
- If not, is he automatically not guilty by reason of insanity?

TESTAMENTARY CAPACITY

To engage in certain activities, one must first be legally qualified to do so. For example, an individual is not eligible to drive a car unless they are at least 16 years old, have good vision, and have passed a driving test. Earlier in this chapter, you read about a different kind of competency: the competency to stand trial. To be competent to stand trial, a person must be able to understand what occurs at court and be able to participate in the legal process.

The law also defines several other competencies that take place outside of the courtroom. These competencies are called *civil competencies*. Some well-known examples include the competency to make a will, to form a contract, and to refuse medical treatment. In each of these examples, a forensic psychiatrist may be asked to evaluate the individual to assess if a mental disorder affects the person's ability to complete the task in question.

The competency to make a will is called *testamentary capacity*. To be mentally capable to prepare a will, the law requires that an individual understand three issues:

the extent of the property being willed, the relatives who ordinarily would receive that property, and the fact that a will is being created. Additionally, the *testator*—that is, the person making the will—must do so voluntarily. A will that is forced or threatened is invalid.

Case Example 5

An 88-year-old widowed man with mild dementia decides to prepare a will after winning several million dollars in the lottery. Estranged from his three children and five grandchildren, the man chooses instead to leave everything to his visiting nurse. "You're the only person who cares about me. Those damn kids haven't called me in years. Why should I give them a penny?" After their father dies, the children contest his will, alleging that they are his rightful heirs and should receive all of the inheritance.

Questions for the forensic psychiatrist:

- Does the man's dementia automatically mean he lacked testamentary capacity?
- Does the fact that the nurse was both his caretaker and beneficiary indicate that she might have pressured him into changing his will?

Sex Offenders

The topic of sex offenders is perhaps the most controversial subject in forensic psychiatry today. Historically, states have been uncertain about how to respond to individuals who commit sex offenses such as rape or child molestation. Do these people suffer from mental illness and require hospitalization? Or are they ordinary criminals who should be incarcerated? Unable to fully resolve these questions, many state legislatures have passed laws requiring both consequences. Convicted sex offenders can first be sent to prison to complete a criminal sentence and then transferred to a forensic psychiatric hospital for treatment. These laws, known, as *sexually violent predator (SVP)* laws, have been passed by approximately one-third of states in the United States since 1990.

Under SVP laws, forensic psychiatrists are typically asked to assess sex offenders at one of two points. The first occurs after the offender has completed their prison sentence, and the court must determine whether to hospitalize them. A forensic psychiatrist is asked to interview the inmate and decide if the individual suffers from a "mental abnormality" that is likely to cause him or her to commit future sexually violent offenses. Each state chooses how to define "mental abnormality," and most do not require the presence of a severe mental disorder. One reason that SVP laws are so controversial is that they allow offenders to be indefinitely hospitalized even without true mental disorders.

The other instance in which forensic psychiatrists are involved in sex offender cases occurs after the individual completes their sentence and is transferred to the hospital. At regular intervals, a forensic psychiatrist evaluates the offender to determine if they would be dangerous if released. This too is very controversial. Studies have shown that experts are only slightly better than chance at predicting future recidivism. Thus, under SVP laws, sex offenders can be hospitalized without mental disorders, effective treatments, or explicit ways to determine when it is safe to release them. However, for some repeat sex offenders who commit violent crimes, SVP laws can be an effective way to protect public safety.

Workers' Compensation

Under civil law, an injured party may sue to obtain financial compensation for damages. In certain circumstances, however, other methods exist to receive payments without bringing a lawsuit. For example, in the early 20th century, special *workers' compensation* laws developed to protect employees injured on the job. These laws eliminate the need for an injured worker to prove that the employer was at fault and ignore the possibility that the employee contributed to the injury. This "no-fault" arrangement avoids costly litigation and ensures that employees will be compensated for lost earnings and medical expenses.

In addition to receiving rewards for physical injury, employees can also seek damages for mental injury. Forensic psychiatrists are frequently consulted when employees claim emotional distress or impairment. A careful evaluation is required to determine alleged symptoms, severity, and causation. Additionally, forensic psychiatrists must assess whether an employee's mental disorder existed prior to the injury. Such a finding may invalidate the employee's claim that symptoms resulted from a work-related accident.

Case Example 6

A middle-aged woman has worked for a large construction company for over 20 years. She has a spotless employment record and has been promoted several times. Recently while at work she witnessed a co-worker fall from a building's scaffolding to his death. For several weeks afterwards, the woman has recurrent nightmares and is unable to sleep. At work, she cries frequently and shakes so badly that she cannot perform her job. She eventually brings a workers' compensation claim against her company alleging posttraumatic stress disorder. A forensic psychiatrist is consulted to evaluate her and learns that, one month prior to the accident at work, her grandson drowned while swimming in her backyard pool. The employee admits that she had already been feeling depressed before the work accident and had been taking psychiatric medication prescribed by her doctor.

Questions for the forensic psychiatrist:

- What work-related injury did the woman sustain?
- Did this employee have a pre-existing mental disorder prior to the accident at work?
- If so, is it possible that this condition was aggravated or accelerated?

Termination of Parental Rights

The last issue of civil law that we will cover is the termination of parental rights. Throughout most of history, the law has treated children as parental property, and children have not had any specific rights of their own. Infanticide, the killing of a newborn child, was tolerated in ancient societies and continues in some parts of the world today. Beginning in the 20th century, however, new laws developed in the United States to protect children's rights. Limitations were placed on child labor, and juvenile delinquents were given many of the same rights as their adult counterparts in criminal court.

Parents continue to have broad discretion when choosing how to raise their children. For example, parents can determine their children's religion, education, and social activities. Within certain boundaries, parents also have the right to discipline their children as they see fit. However, parents do have certain responsibilities to their children. If these obligations are not met, parental rights can be terminated. For example, parental rights can be terminated if a parent abandons, neglects, or abuses a child. Parental rights may also be terminated if a parent has a mental disorder that would put the child at risk for any of these conditions. In these cases, forensic psychiatrists must evaluate the parents and the child to determine the child's "best interests." The decision to terminate parental rights is permanent and complete, and children are subsequently placed in foster care.

Case Example 7

A single, 25-year-old woman gives birth to a healthy baby boy. Over the following two years, she develops a severe, chronic form of schizophrenia that does not respond to treatment. She begins to have the delusion that food is poisonous and stops feeding her son. The boy begins to lose weight and becomes dangerously ill.

Questions for the forensic psychiatrist:

- What are the grounds for the termination of parental rights in this case?
- Before terminating parental rights, what steps could be taken to protect the child?

Hostage Negotiations

As you learned earlier, forensic psychiatrists work in several diverse roles. Some are employed by state prisons to treat mental illness among the inmates. Others serve as consultants to judges and lawyers when questions of mental disorder arise in various criminal or civil lawsuits. Additionally, forensic psychiatrists sometimes work directly with law enforcement personnel. Police officers, for example, interact frequently with people who have mental illness, and they need to be educated about these disorders.

In addition to providing mental health training, forensic psychiatrists can also work with law enforcement to resolve hostage situations. During these crises, the perpetrators use hostages as bargaining tools to demand money, escape, or political objectives. The hostage-takers must keep their victims alive or risk losing their negotiating leverage. The goal of crisis negotiation is to de-escalate the conflict and find nonviolent ways to ensure the safe release of the hostages. National guidelines recommend that negotiation teams consult with a mental health professional, who may be able to provide a unique contribution, such as a psychological profile or risk assessment of the perpetrator. Direct communication with the hostage-takers is usually reserved for law enforcement.

Case Example 8

A 60-year-old man with psychotic symptoms enters a government office brandishing an assault rifle. He demands that the "government stop broadcasting messages into my brain!" and warns that he will start killing hostages unless the voices stop. A standoff ensues for several hours, and the police decide to consult with a forensic psychiatrist.

Questions for the forensic psychiatrist:

- What techniques might be suggested to lower the tension at the scene?
- How could the man be encouraged to surrender?
- Would it be ethical for a forensic psychiatrist to distract the man while a SWAT team secretly moves in to kill him?

Resources for Additional Information

There are many excellent websites where you can learn more about mental illness and about forensic psychiatry. The following are a few suggestions:

The American Academy of Forensic Science (AAFS) www.aafs.org
The American Academy of Psychiatry and the Law (AAPL)
www.aapl.org
The American Psychiatric Association (APA)
www.psych.org
The National Institute of Mental Health (NIMH)
www.nimh.nih.gov
The World Health Organization (WHO)
www.who.int

The author of this chapter is a forensic psychiatrist. You may contact him by
e-mail at DavidTaylorMD@gmail.com.

Quiz

Refer to the text in this chapter if necessary. Answers are located in the back of the
book.

1. Which of the following physicians pioneered the field of *phrenology*, in
   which the contours of patients' scalps were studied to diagnose problems of
   underlying "brain organs"?

   (a) Dr. Franz Gall

   (b) Dr. Sigmund Freud

   (c) Hippocrates

   (d) Dr. Carl Jung

2. According to the NIMH, what percentage of American adults experience
   mental illness in a given year?

   (a) 10%

   (b) 25%

   (c) 50%

   (d) 85%

3. Which of the following is an example of a psychotic symptom?

   (a) Feeling depressed

   (b) Feeling anxious

   (c) Attempting suicide

   (d) Hearing voices

4. Malingering means:

   (a) Acting with malice

   (b) Pretending to be healthy when actually ill

   (c) Pretending to be ill when actually healthy

   (d) Acting with kindness

5. Which of the following is the best description of the adversarial system in American courts?

   (a) Each party retains a lawyer who presents as favorable a version of the truth as possible to the judge or jury.

   (b) Each party hires an expert witness who acts in an adversarial manner.

   (c) The judge is adversarial toward the defendant.

   (d) The jury is adversarial toward the plaintiff.

6. During a sanity evaluation, a forensic psychiatrist evaluates the defendant's _____ mental state.

   (a) Past

   (b) Present

   (c) Future

   (d) All of the above

7. During an evaluation for competency to stand trial, a forensic psychiatrist evaluates the defendant's _____ mental state.

   (a) Past

   (b) Present

   (c) Future

   (d) All of the above

8. Sexually violent predator laws regard sexual offenders as:

   (a) Criminals to be sent to prison

   (b) Mentally ill to be treated in a hospital

   (c) Either a or b

   (d) Both a and b

9. Testamentary capacity refers to the ability to:

    (a) Testify in court

    (b) Give sworn testimony during a deposition

    (c) Complete psychological testing

    (d) Write a will

10. An employee can bring a workers' compensation case if:

    (a) The employer was at fault

    (b) The employee was at fault

    (c) Both the employer and employee were at fault

    (d) Any of the above

APPENDIX

Forensic Science Education, Careers, and Resources

The plethora of TV shows about forensic science has created an interesting phenomenon. High school and college students are considering career paths that will take them into the nation's crime labs as forensic scientists, criminalists, and crime scene investigators. This interest has spawned a large number of university programs that now offer undergraduate and graduate degrees in forensic science. Even traditional academic programs in chemistry and biology are adding forensic science courses to their curricula.

The American Academy of Forensic Sciences' Forensic Science Education Program Accreditation Commission (FEPAC) accredits forensic undergraduate and graduate programs. While FEPAC is a voluntary program, it gives prospective students a way to evaluate quality forensic science education programs.

With the heightened interest in, and the increasing number of students graduating from, these programs, it is only natural to ask the question: what are crime laboratories in the public sector looking for from applicants, and how can students prepare themselves for careers in forensic sciences? Forensic science is applied science. Anyone wishing to pursue a career in forensics needs to be well grounded in science, most especially in chemistry, biology, and physics. The nature of the examinations conducted in forensic science labs is fundamentally science and, as a result, applicants with strong scientific backgrounds are most likely to be successful.

With the large number of applicants flooding crime labs, applicants are coming to interviews with ever stronger academic preparation. Whereas 20 years ago a BS degree would have been all that was needed for that first crime lab job, today, this is no longer the case. Supply and demand has placed crime labs in an interesting position where more and more applicants have MS and PhD degrees in forensic science, chemistry, and biological sciences. But a graduate degree may not be enough to get a job offer.

Many students volunteer at crime labs to gain experience. This has two benefits: it gives the volunteer some hands-on experience, and it gives labs the chance to observe the student to ascertain how he or she will fit in at a given laboratory. An additional benefit for students is that it gives them some important experience to list on a résumé.

Those seeking careers in forensic science are expected to be above-average public speakers. Courtroom testimony is a form of public speaking, and successful applicants can expect to testify in trials. Those who may feel uncomfortable about speaking publicly need to overcome that fear. One suggestion is to join *Toastmasters International* (www.toastmasters.org) to sharpen public speaking skills.

Many forensic science professional organizations have student membership categories. Students should consider joining one or more of these organizations and, if possible, attending meetings. In some instances, students can deliver presentations on their senior projects or graduate research as oral publications or, better yet, submit a paper for publication in journals, such as the *Journal of Forensic Sciences*. There are several regional, national, and international forensic bodies. Some of the major organizations are listed here, all of which can be located on the Web via a search engine:

International

- International Association for Identification (IAI)
- International Association of Forensic Sciences (IAFS)
- European Academy of Forensic Science (EAFS)

National

- American Academy of Forensic Sciences (AAFS)
- American Board of Criminalistics (ABC)
- Australian and New Zealand Forensic Science Society (ANZFSS)
- British Association for Human Identification (BAHID)
- Canadian Society of Forensic Science (CSFS)
- American Board of Forensic Toxicology (ABFT)
- American Board of Forensic Odontology (ABFO)
- National Association of Medical Examiners (NAME)
- Association of Firearm and Toolmark Examiners (AFTE)

Regional (just a few examples)

- Northeastern Association of Forensic Scientists (NEAFS)
- California Association of Criminalists (CAC)
- Mid-Atlantic Association of Forensic Scientists

Students who are applying for jobs in crime labs should prepare themselves before their interviews. Many labs are willing to take prospective applicants through the lab before an interview. Applicants should speak to lab managers and discuss what they are looking for in applicants. A pre-application visit to a lab is a good opportunity to learn about the needs of an organization and to meet the staff on an informal basis.

Similarly, science majors without any forensic experience should research criminalistics and forensic science. How does the chemistry, biology, and physics studied at their university relate to what is done in a crime lab? Applicants who have done little to prepare for their interview often perform poorly at their interviews. Preparation and study are important.

Interviews may be stressful experiences for candidates. This is sometimes done purposefully. Crime lab managers recognize that forensic scientists must be able to handle themselves well in difficult courtroom testimony situations. The interview process hints at the ability of the student to handle his- or herself in difficult situations.

The majority of forensic science positions are in law enforcement agencies that have rigid hiring criteria. Applicants should not be surprised that many require polygraph examinations and background checks as part of the application process. Most police agencies have strict standards concerning drug use. It is necessary to advise an applicant who may have their heart set on working in a crime lab that prior drug use is an impediment to that employment.

Crime labs value employees with high ethical standards. After all, the work conducted in labs may have a direct bearing on the outcome of a trial. Labs and employees of forensic labs have a duty to see that proper science is done in the investigations and subsequent experiments, in addition to ensuring that the scientific data is properly interpreted in reports.

Crime labs come in various sizes, from very large metropolitan labs to small rural labs. Larger labs provide new employees with experience in a relatively short time, whereas rural labs may not see as many unusual cases over the same period of time. On the other hand, larger labs tend to operate within more bureaucratic organizations. Police agencies are semi-military in their approach and quite different from other organizations. Applicants who may have a hard time with a more rigid work environment may feel more at home in smaller labs.

Most crime labs are set up according to two different models. In some labs, forensic scientists do one type of work. For example, forensic DNA analysts may only work in forensic biology for their entire career. Other labs may move personnel from one discipline to another over a career. Another important distinction is lab work versus a combination of lab work and field investigation work. Some labs routinely send forensic scientists to crime scenes and then have them do the lab work, while others do not. If your idea is to be at the crime scene, you should determine what the policy of the lab is where you are applying.

A way to evaluate crime labs is to determine if they are accredited. Accreditation requires labs to meet and maintain quality standards. Two accreditation programs available to labs are the American Society of Crime Laboratory Directors–Laboratory Accreditation Board and the program offered by Forensic Quality Services. Accreditation through either program demonstrates that a lab has met certain standards and produces a quality work product.

Similarly, many labs support individual certification. The ABC and the IAI each offer voluntary certification programs for individual forensic scientists. Some labs use certification as a means to promote individuals. The ABC has begun working with FEPAC to develop a process to evaluate students in forensic science programs.

A career in forensic science is extremely satisfying and rewarding. Few professions offer such an exciting, interesting, and fulfilling career—one that offers scientists a chance to bring justice into their communities and in some small way make our society better.

Final Exam

Select the best answer for the following multiple choice questions. The answers are located at the end of the book.

1. The fundamental idea in forensic science that every contact leaves a trace is known as:

   (a) The conservation of energy

   (b) Locard's Exchange Principle

   (c) The Henry System

   (d) The Combined DNA Indexing System

2. A criminologist is one who studies:

   (a) A crime scene

   (b) Fingerprints

   (c) The *modus operandi* of a criminal

   (d) Trace evidence

3. Which of the following fields is not a subsection of forensic science?

    (a) Entomology

    (b) Engineering

    (c) Odontology

    (d) None; all of the above are subsections of forensic science.

4. In the scientific method, when data is analyzed and compared to the original hypothesis, this is know as:

    (a) The feedback loop

    (b) Individualization

    (c) Locard's Exchange Principle

    (d) The conclusion

5. An item or object that acts as a "known" or standard is an:

    (a) Evidence

    (b) Exclusion

    (c) Element

    (d) Exemplar

6. Who is credited with the first utilization of fingerprints at a crime scene?

    (a) William Herschel

    (b) Henry Goddard

    (c) Paul Kirk

    (d) Henry Faulds

7. Alexander Wiener was working in which office when he discovered the Rhesus factor in human blood typing?

    (a) Center for Disease Control, Atlanta

    (b) Berkley Laboratory for Criminalistics, Berkley, California

    (c) Office of Chief Medical Examiner, New York City

    (d) The Home Office Central Research Establishment, Reading, England

8. The term "genetic fingerprinting" was coined after the work done by:

    (a) Henry Faulds

    (b) Kerry Mullis

    (c) Alec Jeffreys

    (d) Karl Landsteiner

9. Who pioneered the acceptance of document examination in court?

    (a) Paul Kirk

    (b) August Vollmer

    (c) Hans Gross

    (d) Albert Osborn

10. What court case established the judge as the "gatekeeper" for expert witness testimony?

    (a) Daubert v. Merrell Dow Pharmaceuticals

    (b) Frye v. United States

    (c) Kumho Tire Co. v. Carmichael

    (d) None of the above

11. What court case established that standards of admissibility could apply to technical evidence and not merely to scientific evidence?

    (a) Daubert v. Merrell Dow Pharmaceuticals

    (b) Frye v. United States

    (c) Kumho Tire Co. v. Carmichael

    (d) None of the above

12. What court case established the "general acceptance" standard for scientific evidence?

    (a) Daubert v. Merrell Dow Pharmaceuticals

    (b) Frye v. United States

    (c) Kumho Tire Co. v. Carmichael

    (d) None of the above

13. An expert witness:

    (a) Must be a scientist

    (b) Can state their opinion

    (c) Is not bound by the rules of hearsay

    (d) Does not have to undergo voir dire because they are already an expert

14. The trier of fact:

    (a) Can be either a judge or jury

    (b) Must base their verdict on the expert witness's conclusion

    (c) All of the above

    (d) None of the above

15. Fees charged by an expert witness should:

    (a) Be determined by the verdict of a case

    (b) Be paid in advance

    (c) Not be paid, because expert witnesses cannot charge for their services

    (d) Be paid only if the expert reaches the conclusion the defendant wants

16. The statute of limitations:

    (a) Applies to all types of crimes

    (b) Is intended to limit the impact of memories fading over time

    (c) Is suspended once an indictment is brought

    (d) Both b and c

17. To be an expert witness, one needs:

    (a) Education

    (b) Experience

    (c) Certification

    (d) To be qualified by a court

18. The first officer to arrive at the scene of a crime should:

    (a) Cordon off the area surrounding the scene

    (b) Begin immediately collecting evidence

    (c) Provide aid to any living victims

    (d) Both a and c

19. The taking of photographs, notes, and sketches at a crime scene is known as:

    (a) Establishing the chain of custody

    (b) Individualization

    (c) Presumptive testing

    (d) Documentation

20. Good photographs, legible notes, and understandable sketches of a crime scene are important because:

    (a) The detectives on the case might not review the information until much later.

    (b) They will be presented in court and the jury will see them.

    (c) They will be sealed in a plastic coating and cannot be changed later on.

    (d) Both a and b

21. The phenomenon when an investigator becomes focused on only one scenario and ignores other possible scenarios is known as:

    (a) Individualization

    (b) Tunnel vision

    (c) Documentation

    (d) The scientific method

22. The presence of a fine, misted blood spatter pattern at a shooting scene might be an indicator of:

    (a) Low-velocity blood spatter

    (b) Medium-velocity blood spatter

    (c) High-velocity blood spatter

    (d) None of the above

23. The paperwork that accounts for the transfer of evidence from one person or location to another is known as the:

    (a) Voucher

    (b) Laboratory report

    (c) Chain of custody

    (d) Subpoena

24. The method of searching a crime scene in one direction and then again in another direction is known as:

    (a) Spiral search

    (b) Grid search

    (c) Lane search

    (d) Helicopter search

25. Regarding the bitemark photography scale, ABFO stands for:

    (a) Association of Bitemark Forensic Odontologists

    (b) American Bitemark Forensic Organization

    (c) Applied Biosystems Fluorescent Oligonucleotides

    (d) American Board of Forensic Odontology

26. Detail photographs should always be shot at a _____ to the evidence:

    (a) Relatively far distance

    (b) 45-degree angle

    (c) 90-degree angle

    (d) 10-degree angle

27. The type of film that is best for documenting contrast differences is:

    (a) Color film

    (b) Monochrome film

    (c) Infrared film

    (d) Polaroid film

28. The circular ring in a camera lens that can be opened or closed to control the amount of light is known as the:

    (a) Focusing ring

    (b) Shutter

    (c) Aperture

    (d) CCD

29. The SWGIT recommendations for digital photography:

    (a) Discourage the use of digital cameras at a crime scene

    (b) Recommend that all photos be scanned into computers for image manipulation

    (c) Establish guidelines for the use, presentation, and archiving of digital images

    (d) Are generally ignored by the forensic science community

30. The process in which a body is examined internally and externally to find the cause and manner of death is known as the:

    (a) Procedure

    (b) Identification

    (c) Autopsy

    (d) None of the above

31. Determining whether a death is a homicide, suicide, natural, accidental, or unknown is establishing the _____ of death.

    (a) Mode

    (b) Method

    (c) Means

    (d) Manner

32. *Algor mortis* and *rigor mortis* can aid a pathologist in determining which of the following?

    (a) Mode of death

    (b) Time of death

    (c) Identification of the body

    (d) Individualization of the body

33. A forensic pathologist might be called in on which of the following cases?

    (a) A fatal shooting

    (b) An apparent drug overdose

    (c) A woman who was beaten by her husband and survived

    (d) All of the above

34. During an autopsy, the medical examiner is looking for:

    (a) External wounds

    (b) Internal trauma

    (c) Disease

    (d) All of the above

35. Examination of the stomach contents during an autopsy can reveal:

    (a) The final meal of the deceased

    (b) The time of death

    (c) The bubblegum the deceased swallowed 10 years prior to death

    (d) Both a and b

36. The forensic study of bones and skeletonized remains is known as:

    (a) Osteology

    (b) Anthropomorpy

    (c) Anthropology

    (d) Entomology

37. The most commonly used tool in trace evidence analysis is a:

    (a) Shovel

    (b) HPLC

    (c) Bone saw

    (d) Forceps

38. Trace evidence can only be analyzed if it is:

    (a) Visible to the naked eye

    (b) Requested by the detective in the case

    (c) Located and preserved

    (d) Capable of exonerating the suspect

39. The process in which light is broken into its distinct colors when passing through a lens is known as:

    (a) Spherical aberration

    (b) Köhler illumination

    (c) Darkfield microscopy

    (d) Chromatic aberration

40. The part of the microscope that the user looks into to see the image is known as the:

    (a) Ocular

    (b) Objective

    (c) Field diaphragm

    (d) Condenser

41. When two polarizing filters are placed 90 degrees to each other, this is known as:

    (a) Darkfield microscopy

    (b) Refraction

    (c) Acute total reflectance

    (d) Crossed polars

42. Fourier transform is a _____ conversion.

    (a) Theoretical

    (b) Physical

    (c) Mathematical

    (d) Religious

43. The three growth phases of hair are:

    (a) Adenine, cytosine, thymine

    (b) Actin, cortisone, Tylenol

    (c) Anagen, catagen, telogen

    (d) Anisotropic, cation, telomerase

44. The oil glands on a finger are located:

    (a) On the pad of the finger

    (b) On the sides of the finger

    (c) Under the nail of the finger

    (d) Fingers do not contain oil glands.

45. Reference sets in a trace evidence laboratory should be made for which of the following:

    (a) Animal hairs

    (b) Plant fibers indigenous to that region

    (c) Common synthetic fibers

    (d) All of the above

46. Alpaca fibers are obtained from which animal's hair?

    (a) Cattle

    (b) Llama

    (c) Chinchilla

    (d) Rabbit

47. Glass containing borax is marketed under the trade name:

    (a) Leucite

    (b) Plexiglas

    (c) Avanit

    (d) Pyrex

48. In the United States, normal tempered glass can be found:

    (a) On the driver's side windows of a car

    (b) On the passenger's side windows of a car

    (c) In the windshield of a car

    (d) Only a and b

49. Linen fibers are made from the stem of which plant?

    (a) Flax

    (b) Cotton

    (c) Hemp

    (d) Juniper

50. In a polarized light microscope, the secondary polarizing filter is known as the:

    (a) Compensator

    (b) Analyzer

    (c) Micrometer

    (d) Condenser

51. In microscopy, when the uniformly illuminated lamp lens is framed by the field diaphragm and produces an ideal color temperature, what is this known as?

    (a) Köhler illumination

    (b) Polarized light microscopy

    (c) Comparison microscopy

    (d) Oblique illumination

52. A _____ test is used to screen a sample and might result in a false positive.

    (a) Confirmatory

    (b) Individualizing

    (c) Serology

    (d) Presumptive

53. Following a rape attack, and sexual assault evidence kit should be collected:

    (a) After the victim has a chance to shower

    (b) Immediately

    (c) Only after a suspect has been arrested

    (d) A sexual assault kit is not collected after rape attacks.

54. When the same DNA alleles are present on each chromosome, this is known as:

    (a) Homozygous

    (b) Heterozygous

    (c) Heteroplasmy

    (d) A short tandem repeat

55. In DNA testing, buccal swabs are another term for:

    (a) Vaginal swabs

    (b) Anal swabs

    (c) Oral swabs

    (d) Bloodstain swabs

56. Epithelial cells are present in which of the following body locations?

    (a) Inside of the mouth

    (b) On the surface of the skin

    (c) In the walls of the vagina

    (d) All of the above

57. DNA can be found in:

    (a) The nucleus of the cell

    (b) The cell wall

    (c) The mitochondria within a cell

    (d) Both a and c

58. PCR is a DNA replication process that uses _____, chemicals, polymerase, and template DNA.

    (a) Primers

    (b) Exonucleases

    (c) Nitriles

    (d) Locard's Exchange Principle

59. The final graphical data that is used in DNA analysis is known as the:

    (a) Quanitblot

    (b) Polymerase chain reaction

    (c) Electropherogram

    (d) Electrophoresis

60. Which of the following people would have a Y STR profile?

    (a) Mother

    (b) Daughter

    (c) Uncle

    (d) Niece

61. The mitochondria in a sperm cell are located:

    (a) In the head

    (b) In the neck

    (c) In the tail

    (d) Sperm do not have mitochondria.

62. The science of the physics behind a moving projectile is know as:

    (a) Acoustics

    (b) Buoyancy

    (c) Ballistics

    (d) Reflectance

63. The caliber of a bullet is a measure of the:

    (a) Length of the bullet

    (b) Weight of the bullet

    (c) Diameter of the bullet

    (d) Circumference of the bullet

64. Which of the following weapons contains a "cylinder gap?"

    (a) Semi-automatic pistol

    (b) Revolver

    (c) Rifle

    (d) Shotgun

65. The rifling in a shotgun barrel is produced by:

    (a) Hammering

    (b) Cutting

    (c) Molding

    (d) Shotguns do not have rifled barrels.

66. The primer in ammunition is:

    (a) Unstable

    (b) A low explosive

    (c) A high explosive

    (d) Both a and c

67. The _____ removes a cartridge from the chamber of a weapon.

    (a) Breech

    (b) Extractor arm

    (c) Magazine

    (d) Hammer

68. The optical instrument used to observe two items simultaneously is known as:

    (a) A compound microscope

    (b) An electron microscope

    (c) A stereo microscope

    (d) A comparison microscope

69. Which federal bureau maintains the NIBIN database?

    (a) FBI

    (b) CIA

    (c) ATF

    (d) DEA

70. The best chemical to use when collecting GSR from a suspect's hands is:

    (a) 50% hydrochloric acid

    (b) Water

    (c) Dilute nitric acid

    (d) Concentrated sulfuric acid

71. The Greiss test screens for the presence of:

    (a) Sulfur

    (b) Nitrites

    (c) Blood

    (d) Cocaine

72. The chemical reagent that is used to eat away uncompressed metal is:
    (a) Kastle-Meyer reagent
    (b) Acid phosphatase
    (c) Sodium rhodizonate
    (d) Fry's reagent

73. Which of the following is an example of impression evidence?
    (a) A bitemark into a block of cheese
    (b) The firing pin mark on the back of a round of ammunition
    (c) The text on a ransom note typed on a typewriter
    (d) All of the above

74. Which of the following is not a separation technique?
    (a) TLC
    (b) HPLC
    (c) GC-MS
    (d) NMR

75. Drugs that have a low potential for abuse and a high therapeutic value are:
    (a) Class III
    (b) Class I
    (c) Class IV
    (d) Class V

76. Which of the following is not an inorganic poison?
    (a) Thallium
    (b) Arsenic
    (c) Digoxin
    (d) Lead

77. A Quaalude is a(n):
    (a) Benzodiazepine
    (b) Opiate
    (c) Barbiturate
    (d) Amphetamine

78. TLC is:

 (a) A separation technique

 (b) A spectroscopic technique

 (c) A qualitative technique

 (d) Both a and c

79. GC uses which type of mobile phase?

 (a) Gas

 (b) Silica

 (c) Plates

 (d) Liquid

80. Which of the following is a depressant?

 (a) Phenobarbital

 (b) Ethanol

 (c) Valium

 (d) All of the above

81. Crack is produced by boiling cocaine with:

 (a) Baking powder

 (b) Salt

 (c) Baking yeast

 (d) Baking soda

82. A substance that is used to speed up or increase the heat of a fire is known as a(n):

 (a) Accelerant

 (b) Retardant

 (c) Explosive

 (d) Detonator

83. The first step in arson reconstruction is to locate the _____ of the fire.

 (a) Cause

 (b) Method

 (c) Origin

 (d) Termination

84. Arson evidence needs to be sealed in metal canisters, due to:

    (a) The fragility of the evidence

    (b) The dampness of the evidence

    (c) The volatility of common accelerants

    (d) The explosive potential of the evidence

85. When testing arson evidence, the method of puncturing a canister with a syringe and directly sampling the air in the canister is known as:

    (a) Passive concentration

    (b) Simple headspace

    (c) Dynamic concentration

    (d) Solvent extraction

86. The process in which an explosive burns rather than expands is known as:

    (a) Fragmentation

    (b) Implosion

    (c) Deflagration

    (d) Arson

87. The lethal effect of an explosion is:

    (a) Heat

    (b) The shockwave

    (c) Shrapnel

    (d) All of the above

88. A good forensic document examiner must not be:

    (a) Unbiased

    (b) Experienced

    (c) Colorblind

    (d) Certified

89. The American Board of Forensic Document Examiners only issues certification after:

    (a) Successful completion of the written, practical, and oral exams

    (b) Completion of a two-year apprenticeship under a senior scientist

    (c) Completion of two years of independent casework

    (d) All of the above

90. Differences between two inks on a document can best be "seen" with:

    (a) FTIR instrumental analysis of the inks

    (b) TLC separation of the inks

    (c) GC-MS analysis of the inks

    (d) Infrared photography of the document

91. Which reagent is best for developing fingerprints on paper?

    (a) Kastle-Meyer

    (b) Acid phosphatase

    (c) Sodium rhodizonate

    (d) Ninhydrin

92. ESDA is an acronym for:

    (a) Eastern Society of Document Analysts

    (b) Early screening for document approval

    (c) Electrostatic detection apparatus

    (d) Erstwhile sanctions on document applications

93. The process in which a document is ensured to not be a forgery is:

    (a) Certification

    (b) Accreditation

    (c) Verification

    (d) Authentication

94. Which of the following countermeasures are not present on U.S. currency?

    (a) Security threads

    (b) Color-shifting ink

    (c) Traceable nanoparticles

    (d) Microprinting

95. Modern perspectives on mental illness are approached from a model known as:

    (a) Phrenology

    (b) Psychoanalysis

    (c) Biopsychosocial

    (d) Shock therapy

96. The mental illness of "split personalities" is known as:

    (a) Schizophrenia

    (b) Bulimia

    (c) Dissociative identity disorder

    (d) PTSD

97. The process in which a person fakes a mental illness is known as:

    (a) Malingering

    (b) Pleurisy

    (c) Cognition

    (d) Hearsay

98. A forensic psychologist can determine:

    (a) If a person is competent to stand trial

    (b) If a person is criminally insane

    (c) If a person is guilty of a crime

    (d) If a person is lying

99. A forensic psychologist might be called in to deal with which of the following scenarios?

    (a) A workers' compensation investigation

    (b) A hostage negotiation

    (c) A case involving the welfare of a child

    (d) All of the above

100. The AAFS FEPAC program:

    (a) Issues certification to individual criminalists

    (b) Grants waivers to nonaccredited crime laboratories

    (c) Nominates candidates for the Paul Kirk Award

    (d) Provides certification to undergraduate and graduate forensic science programs

Answers to Quiz and Exam Questions

Chapter 1: Introduction

1. C	2. B	3. C	4. D	5. C
6. B	7. B	8. A	9. C	10. B

Chapter 2: Legal

1. B	2. D	3. A	4. C	5. C
6. D	7. D	8. B	9. A	10. D

Chapter 3: Crime Scene Investigation

1. C	2. C	3. A	4. B	5. D
6. D	7. D	8. D	9. D	10. B

Chapter 4: Forensic Photography and Documentation

1. C	2. B	3. A	4. A	5. D
6. B	7. C	8. D	9. A	10. D

Chapter 5: Pathology

1. C	2. D	3. C	4. C	5. D
6. C	7. B	8. B	9. C	10. D

Chapter 6: Trace Evidence

1. D	2. A	3. C	4. B	5. D
6. C	7. C	8. B	9. A	10. B

Chapter 7: Forensic Biology/DNA

1. C	2. C	3. D	4. D	5. C
6. B	7. D	8. D	9. D	10. D

Chapter 8: Firearms and Toolmarks

1. D	2. A	3. B	4. D	5. D
6. D	7. A	8. D	9. D	10. C

Chapter 9: Drugs and Toxicology

1. D	2. D	3. D	4. D	5. B
6. D	7. A	8. A	9. A	10. D

Chapter 10: Arson and Explosives

1. D	2. D	3. D	4. D	5. C
6. C	7. B	8. D	9. A	10. C

Chapter 11: Questioned Documents

1. B	2. C	3. D	4. B	5. D
6. D	7. B	8. D	9. D	10. A

Chapter 12: Forensic Psychiatry

1. A	2. B	3. D	4. C	5. A
6. A	7. B	8. D	9. D	10. D

Final Exam

1. B	2. C	3. D	4. A	5. D
6. D	7. C	8. C	9. D	10. A
11. C	12. B	13. B	14. A	15. B
16. D	17. D	18. D	19. D	20. D
21. B	22. C	23. C	24. B	25. D
26. C	27. B	28. C	29. C	30. C
31. D	32. B	33. D	34. D	35. D

36. C	37. D	38. C	39. D	40. A
41. D	42. C	43. C	44. D	45. D
46. B	47. D	48. D	49. A	50. B
51. A	52. D	53. B	54. A	55. C
56. D	57. D	58. A	59. C	60. C
61. B	62. C	63. C	64. B	65. D
66. C	67. B	68. D	69. C	70. C
71. B	72. D	73. D	74. D	75. D
76. C	77. C	78. D	79. A	80. D
81. D	82. A	83. C	84. C	85. B
86. C	87. D	88. C	89. D	90. D
91. D	92. C	93. D	94. C	95. C
96. C	97. A	98. A	99. D	100. D

INDEX